The Realm of the Terrestrial Planets

The New Frontier—Rendezvous of Man and Machines on the Moon
On 29 November 1969, astronauts of the Apollo 12 mission left their landing module *Intrepid* (on the horizon) to visit and retrieve part of the equipment of Surveyor 3 (foreground). When the sealed compartment of the TV system aboard Surveyor 3 was brought back to Earth, it was discovered that a virus of the common cold had found its way into the compartment and was still alive after a 31 month forced exposure to the lunar environment—no mean feat of survival! *Photograph by courtesy of NASA and Hasselblad.*

The Realm of the Terrestrial Planets

Zdeněk Kopal

The Institute of Physics
Bristol and London

Copyright © 1979 Zdeněk Kopal

Published by The Institute of Physics
Techno House, Redcliffe Way, Bristol BS1 6NX, and
47 Belgrave Square, London SW1X 8QX

British Library Cataloguing in Publication Data

Kopal, Zdeněk
    The realm of the terrestrial planets.
    1. Planets
    I. Title   II. Institute of Physics
    523.4   QB601
    ISBN 0-85498-034-2

Filmset in 10/12 point Times New Roman and 8/10 point Univers
Printed and bound in Great Britain
by W & J Mackay Limited, Chatham, Kent

To Harold Urey
pioneer in the investigation
of the solar system

Preface

The aim of this volume will be to provide — within the limits dictated by its size — an outline of the present extent of our acquaintance with our nearest celestial neighbours: the Moon, the terrestrial planets, and that relatively small corner of the solar system which happens to be our cosmic home. Until only a few years ago we possessed merely a nodding acquaintance with its denizens, since observations could only be made at a distance with the aid of telescopes. The dramatic emergence of long-range spacecraft, which were able to shake off the gravitational bonds of the Earth to venture into the ever-increasing depths of space, has altered the old time-honoured picture almost beyond recognition. In September 1959 the first of these spacecraft landed on the Moon as a harbinger of the manned missions of 1969–1972. And it was not only the Moon that opened unstintingly its ancient treasure chest. One by one, the neighbouring planets also began to yield their secrets to the inquisitive eyes of the nimble-footed intruders sent out from the Earth to reconnoitre them at ever-diminishing distances. And those intruders which in 1959 unveiled the far side of the Moon to our eyes are now penetrating the veil of clouds which surrounds the face of our sister-planet Venus.

All this adds up to a very exciting story which clamours to be told. Unfortunately, since none of the space missions to the other planets has yet found any trace of life there, this story can be addressed only to our fellow-inhabitants of the Earth, who may wish to spare some hours of their time to join us in a guided tour of our immediate cosmic environment. It will be an exciting story — one which our forefathers would have found incredible; it would sound incredible to us if it were not based on undeniable evidence. To follow the factual part of the story should not be difficult. We have endeavoured to make it intelligible to any student of the subject by the omission of superfluous technicalities. But it must be prefaced by a word of caution: namely, the contents of this book should be regarded only as the opening part of the story, the end of which is nowhere yet in sight.

Zdeněk Kopal

Contents

Preface vii

Introduction   The Solar System: The New Frontier   1

1   The World of the Planets: Its Design and Inhabitants   7

2   Exploration by Spacecraft   13
    Planetary Space Probes

3   The Moon: Our Nearest Celestial Neighbour   39
    Motion of the Moon: Months and Eclipses
    Physical Properties of the Moon: Rotation, Size and Mass
    Internal Structure of the Moon
    The Surface of the Moon and its Formations
    Cratering of the Lunar Surface
    The Lunar Environment
    Structure and Composition of the Lunar Surface
    Chronology of the Lunar Surface
    Origin and History of the Moon

4   Mercury and Pluto: The Sentinels of the Solar System   99

5   Mars: The Portrait of a Midi-Planet   111
    Facts and Figures
    The Martian Environment: Its Climate and Atmosphere
    The Mariners and Vikings

The Surface of Mars
Craters on Mars
The Interior of Mars
Martian Satellites

6 Micro-Planets: Asteroids and Lesser Denizens of Interplanetary Space — 151
Asteroids
Asteroids: Dimensions, Mass and Rotation
Meteorites and Shooting Stars
Interplanetary Dust: Zodiacal Light and Gegenschein

7 Venus: The Veiled Planet — 173
Exploration by Radar; Axial Rotation
The Atmosphere of Venus
The Veil of Clouds
Surface Structure and Internal Constitution
Venus: The Mystery Wrapped in an Enigma

8 Our Earth: The Queen of the Terrestrial Planets — 195
Vital Statistics: Motion, Size and Mass
Internal Structure of the Terrestrial Globe
The Crust of the Earth and its Properties:
    Hydrosphere and Atmosphere
History of the Earth and its Future

Index — 219

Introduction
The Solar System: The New Frontier

Perhaps as a heritage from his nomadic past, it has always been man's innate urge to see what lies beyond the horizon of his local habitat; and as more organized ways of travel gradually replaced legs as the only means of locomotion, this horizon began to widen. It was the lure of foreign lands and uncharted seas that led men to seek adventure in distant parts of the world. As the means of travel and communications improved, the originally isolated pockets of ancient civilizations began to coalesce into wider communities.

Two thousand years ago or more, the world of our ancestors extended from the Pillars of Hercules to the Islands of Cathay and from the mists of Ultima Thule to the torrid zone of Africa. The great oceans west of Cadiz and east of the lands of the Grand Khan remained uncharted, and how far south the coasts of Africa may have been followed by intrepid sailors from the Mediterranean basin is still a matter of conjecture. But the sky above remained inaccessible—a closed domain at an unspecified height, inhabited only by gods.

Yet the relentless march of progress did not stop at this stage. In looking over the past history of man's efforts to become acquainted with his surroundings, we cannot fail to note that his horizons have not been widening at a uniform rate, but that long quiescent periods have been interrupted now and then by sudden bursts of activity. This was true, for instance, within the time between the first voyage of Christopher Columbus in 1492 and the first circumnavigation of the Earth by Fernandez Magellan in 1519–1521. In the period between these events, the previously uncharted expanses of the world—two-thirds of its surface in longitude—were explored by brave sailors whose efforts resulted in the surprise discovery of a new continent. The spherical shape of the Earth, which was previously entertained only as a hypothesis by abstruse scholars, then became comprehensible to the man in the street.

The voyages of discovery undertaken in the course of the sixteenth and subsequent centuries do not belong within the scope of this book, and only a brief mention of them will be made here. Ships sailed, arms clashed, and blood flowed in that confrontation between human societies at different levels of development, as a necessary price to pay for the widening of our geographical horizons; by guile and gunpowder, treason and intrigue, motivated by courage and cupidity, the early conquistadores claimed more and more distant lands for their monarchs and their Church. A cosmographer would merely note that, since the discovery of the Antarctic continent in 1772, our knowledge of the principal features of the terrestrial geography had become essentially complete; the only task for the future was to fill in the details.

A hundred years ago, only a few regions of the terrestrial surface remained still untrodden by human foot, and by the beginning of the twentieth century even these had dwindled into insignificance. Since that time, the poles of our planet have been attained and the highest mountains scaled; even the abyssal depths of the oceans have been plumbed and stratospheric altitudes in the atmosphere reached by the descendants of Icarus who, for the first time, beheld the Earth as a sphere. As the remaining terrestrial frontiers were conquered one by one by man's daring and perseverance, another frontier was gradually opening up on other worlds beyond the Earth, particularly on the Moon.

Man's efforts to become acquainted—in fact or fancy—with these new worlds antedate the discovery of the telescope in the early years of the seventeenth century, but, in the hands of astronomers, it is this instrument which has shown the way. Under the searching eye of this new optical device which Galileo Galilei first turned to the sky in the autumn months of 1609, unbridled imagination began to give way to sober fact. At the beginning of the first century of telescopic astronomy Johannes Kepler established the true shape of planetary orbits, and by the end of that century the first realistic mensuration of their sizes, as well as of the planetary globes describing them, had been recorded. It was then that our ancestors learned about the size and (some) physical properties of these new worlds, and also about the distances one would have to travel to reach them.

Until the middle of the twentieth century such travels could take place only in our imagination. The planetary worlds and their secrets beyond the limits of telescopic resolution seemed safe in their inaccessibility, and remained fair prey to science fiction writers. Yet, as time went on, an increasing number of bolder individuals began to turn their eyes to the sky in the spirit which once made Columbus see America from the shores of Spain. As our ground-based telescopes would not disclose (because of atmospheric, rather than instrumental, limitations) all we wanted to know, more and more people became inclined to go and see for themselves.

Earlier in these introductory remarks we mentioned that human knowledge does not grow uniformly but rather in sudden bursts of activity, just like a river sometimes meandering slowly through plains and now and then suddenly accelerated in rapids. Such an acceleration in our knowledge of the Earth occurred between 1491 and 1521—the time between Columbus and Magellan—when our planet was made to disclose so many of its geographical secrets. In the sky, an equivalent acceleration took place after the discovery of the telescope in the early seventeenth century, and again in the latter part of the eighteenth century when William Herschel 'broke the barriers of the heavens' with the aid of his large reflectors. However, none of these events can compare with what we have seen in our lifetime—the advent of the space age.

The space age burst upon us almost unexpectedly. And for good reason: it took three independent branches of human science and engineering—rocket propulsion, long-range radio transmission and computer control—to translate the bold dreams of the early pioneers into serviceable hardware which enabled us to lift scientific instruments, and eventually men, beyond the confines of our atmosphere into the increasing depths of space. With the advent of spacecraft, in a real sense the sky ceased to be the limit. Barely two years elapsed between the first sputnik of October 1957 and Luna 2, which transferred the first particles of terrestrial matter to the lunar surface on 13 September 1959. Only a little more than seven years elapsed between the first manned flight around the Earth by Yuri Gagarin on 12 April 1961 and the first circumnavigation of the Moon by the crew of Apollo 8 on 24 December 1968.

Christmas Eve 1968 became another memorable date for mankind: on that day *Homo sapiens* left his terrestrial cradle for the first time and embarked on his proliferation throughout the solar system. For the 20 hours during which the three American astronauts, F Borman, J A Lovell and W A Anders, circled the Moon, they became its temporary denizens in the same sense as those astronauts who circled the Earth were still terrestrials. In May 1969 the time spent in voluntary lunar captivity was extended to three days by the astronauts of the Apollo 10 mission. It was, however, not until 20 July of the same year that the *Eagle* of the Apollo 11 mission descended to the lunar surface to take, in the words of the Mission Commander, Neil A Armstrong, 'one small step for a man, but a giant leap for mankind' (see plate 1).

Since that time, 27 men have spent 302 hours on the lunar surface or in lunar orbit. Although this first chapter in the era of manned exploration of our satellite is at least temporarily over, unmanned exploration of other celestial targets has continued unabated. A scenario of the domain of this exploration will be outlined in Chapter 1 and some of the space probes will be described in Chapter 2; the rest of the book will present the main results obtained largely with these aids.

It should be stressed here that between 1957, the year of the first sputnik, and 1977, when this book was written, exploration of the Universe by means of space probes has altered our picture of at least the inner precincts of the solar system almost beyond recognition. What were once remote celestial bodies accessible only by telescope are now abodes made familiar to us by television cameras which filmed them at close range, by instruments which landed on their surfaces, and by men who returned to tell

Plate 1. 'One small step for a man, but a giant leap for mankind.' The first man on the Moon, the Apollo 11 Mission Commander, Neil A Armstrong, about to descend to the lunar surface on 20 July 1969.

their story. One aspect of this historical exploration deserves to be emphasized: although the combined total mileage of all the manned missions to the Moon between 1968 and 1972 exceeded 190 million man-kilometres, no life was lost, or injury sustained, by any participant—even the near mishap of Apollo 13 in April 1970 ended happily with the return of the astronauts to Earth. This enviable record has established that space travel is the safest form of travel known to man so far! What a contrast with the risks accepted in sea travel at the time of the voyages of discovery on the Earth, or with those taken by the pioneers of air flight in times closer to our own!

As a result of the recent advances of space travel, the new frontier has now been pushed beyond the orbit of Mars. Even as this book is being written, no less than four automatic probes are en route through space to reconnoitre the outer parts of the solar system beyond the orb of Jupiter.

While the lure of wider horizons will no doubt continue to exert a powerful attraction for future explorers, the planets and other bodies populating the inner precincts of our solar system have probably already yielded the main part of their secrets. It may, therefore, be timely to summarize these in non-technical language for the more general reader, in a manner which future work is unlikely to alter in any essential respect. It is to this end that the present book was written. Most of the data on which the narrative will be based are barely 20 years old, and much of them were obtained in the past 10 years or less: the first landing on the Moon was staged by the Apollo 11 spacecraft in 1969; and the most important planetary missions have been undertaken only since 1970.

The quest for pure knowledge was, in fact, not the only reason these harbingers of new times were sent on their far-reaching journeys. They also served as the pioneers and test vehicles of an entirely new branch of human technology capable of many applications—terrestrial as well as celestial—and their contributions to pure science have been inestimable. Any account of the subject written before their advent would read like medieval Latin or Greek in comparison! Much of the previous guesswork or speculative arguments, so characteristic of the earlier literature, can now be replaced by solid facts, and in many cases these have turned out to be much more exciting than any amount of science fiction.

This does not mean, of course, that all our knowledge of the solar system dates back from yesterday. Centuries—nay, millennia—of patient effort preceding the space age were needed to unravel the master plan of this celestial clockwork and its scale. This is true, in particular, of the motions of the Moon or of the planets. Therefore Chapters 3–8, which deal with the individual members of our planetary family, will open with an account of their kinematic properties which are fundamental for a proper understanding of the more sophisticated parts of our story. In due course the latter will take us to the front line of current research, a domain which must be labelled: 'Men at Work—Pass at Your Own Risk'.

We shall strive to minimize these risks by resisting a temptation to speculate freely on problems too hypothetical to give us confidence that we are on the right track. But we shall not leave out altogether such speculations from our narrative, for we are only too well aware that the unknown points of a problem may constitute part of its attraction. We promise not to cut corners needlessly, nor camouflage ignorance by scenarios of our own making. We must be mindful of the fact that the outcome of every space mission of the past has necessitated many revisions of previously held views; and while this process is decelerating, it is unlikely that it has come to an end.

1 The World of the Planets: Its Design and Inhabitants

In embarking on this brief account of the solar system and of our position within it, it is only appropriate that we should start with a few words on what is by far the most important member of it—the Sun. We discuss the Sun now because we shall not have much opportunity to do so later. Although its overwhelming mass provides the gravitational anchor around which everything else in its neighbourhood revolves, and its radiation provides most of the energy for the activation of many processes which have been shaping the planetary surfaces from time immemorial, we are only too accustomed to take the Sun for granted and, indeed, shall do so throughout most of this book.

The principal reason for the preponderant role of the Sun in our system is its overwhelming mass and size; its luminosity (i.e. its ability to extract radiant energy from its mass) is a direct consequence of these factors. If we compare the Sun with the Earth we find it to be 109 times as large in diameter and 333 000 times as massive. It is the gravitational attraction of this enormous mass which swings the Earth around the Sun in a period of one year at a distance of 149·6 million kilometres from the Sun (a distance hereafter referred to as an 'astronomical unit' of length, or AU for short). The same mass similarly controls the orbits of all the other planets and also guides spacecraft sent out from the Earth on their journeys of exploration through interplanetary space.

This overpowering position of the Sun had already been realized by our intellectual ancestors of the ancient world. Aristarchos of Samos deduced it (from the duration of different eclipse phenomena) as far back as the third century BC, though its dominant role was not always accepted by subsequent generations. 'Which is more useful, the Sun or the Moon?' inquired Kuzma Prutkov, a delightful character created by Alexei Tolstoy; and he answered, 'The Moon, because it shines at night, when it is dark; while the Sun shines only in daytime.'

Of course, every schoolboy knows today that moonlight (on which we shall have more to say later) is only reflected or transformed sunlight. And what is true of the Moon is also true of the planets. When the Sun—whose light dominates our days—sets below the horizon and darkness descends on a tired world ready to seek rest in sleep, most of the celestial objects we can see in the sky are stars. These are objects akin to our Sun, some intrinsically brighter but most of them emitting less light, and they appear so faint because they are situated at distances from which light travelling at 300 000 km s^{-1} will need decades, or centuries, to reach us. In fact, the principal characteristic of the space surrounding us is its overwhelming emptiness. Of all the stars that we can see at night with the naked eye in both hemispheres of the sky, only nine are nearer to us than 15 light years. Although almost 40 more stars are known to lie within this distance, most of them are so faint that large telescopes are needed to make them visible.

Since time immemorial, the starry sky has seemed to present us with an epitome of cosmic immutability. This picture has been modified somewhat in the past few hundred years by the discovery that the stars are in perpetual motion, which appears to us to be so slow only because we measure it over time-spans comparable with those of human life. Since Palaeolithic times the appearance of the sky has certainly changed many times beyond recognition.

However, not all celestial objects visible to the naked eye appear to maintain the same positions in the sky from year to year, or even from month to month. The perennial wanderings of the Moon—the only other celestial object besides the Sun which exhibits a finite diameter to the eye—have attracted man's attention throughout his history. But people of the earliest civilizations in Egypt or Mesopotamia had already noted that some of the brightest 'stars' also wander in the sky in a complicated manner, and their celestial itineraries have fascinated the human mind ever since. For a long time the number of these 'wandering stars' remained at five, and these are best known to us under their Latin names of Mercury, Venus, Mars, Jupiter and Saturn. The first two are visible alternately in the morning and evening sky, while the last three can remain visible throughout the night.

The apparent motions of these 'wandering stars'—or *planets* as we call them—were for many centuries interpreted as being caused by their revolutions around the Earth, though machinery of an increasing degree of complexity (and powered by supranatural causes) had to be postulated to account for their observed celestial itineraries and timetables. It is true that the idea of a *heliocentric* solar system with the Sun at the centre and all planets (including the Earth) revolving around it had already been propounded by Aristarchos of Samos in the third century BC. (His work on the subject is, unfortunately, lost; we know about it only through references made to it in a contemporary tract by Archimedes.) But the idea of a heliocentric system flickered briefly through that first 'century of genius', to be discarded into oblivion for many centuries to come. It was, however, never completely forgotten, and eventually it was revived because of a growing exasperation with its geocentric alternatives. In the first part of the sixteenth century, towards the twilight of the Renaissance, Nicolaus Copernicus (1473–1543) revived the heliocentric planetary system—albeit without much success—by dressing it up in the garb of Ptolemaic epicycles. But it took the discovery of the telescope and a major theological revolution in the first half of the seventeenth century to establish the validity of the heliocentric system and gain for it a general acceptance.

This last accomplishment was in fact not due to Copernicus, but to his spiritual heirs Galileo Galilei (1564–1642) and Johannes Kepler (1571–1630). It was Kepler who swept away the cobwebs of medieval astronomy from the sky by recognizing the planetary orbits for what they are: namely, ellipses, with the Sun at their focus, and the planets revolving around the Sun with an angular velocity inversely proportional to the square of their distance. He thus removed the 'epistemological trash' of the ancients, according to which all celestial motions, being perfect, had *a priori* to be uniform and circular.

It is true that Kepler deduced his laws of planetary motion empirically from the observations made by Tycho Brahe (1546–1601), the last great observer of the pre-telescopic era. The reason why planets move in this way had to await the advent of Isaac Newton (1642–1727) and his 'celestial mechanics' based on the inverse-square law of universal gravitation, according to which the attraction of two bodies varies in proportion to the product of their masses and in inverse proportion to the square of their distance. Kepler himself stumbled upon the same law, heuristically, 80 years before Newton, but was unable to prove that his laws of planetary motion were necessary consequences of the inverse-square law. To do so required a knowledge of the calculus created by Leibnitz and Newton in the closing decades of the seventeenth century; in Kepler's time, this was still 80 years in the future.

Moreover, although the correct geometrical model of the solar system was derived by Kepler from Tycho Brahe's observations in the first decade of the seventeenth century, a realistic idea of its true scale did not dawn upon an astonished humanity until the end of that century, when its magnitude was established by triangulation of the distance of Mars by the French academicians between 1672 and 1684. The true dimensions of the Sun itself, and of the Earth and the other planets, came to light at the same time. This astronomical revolution of the seventeenth century saw the Earth finally dethroned from its assumed privileged position in the Universe to become the sixth planet. After the advent of telescopic astronomy, the years 1781, 1846 and 1930 brought further discoveries of three additional planets beyond Saturn—Uranus, Neptune and Pluto—to complete the planetary system as we know it today. Uranus is only just visible to the unaided human eye, and Neptune or Pluto require telescopes of increasing power for their detection.

The solar system consists of many more objects than the central star with its nine planets. Several of these planets are attended by families of *satellites* of their own and two (Saturn and Uranus) are girdled with *rings* made up of an untold number of small particles revolving around the planet. Moreover, the Sun itself is surrounded by a loose ring of 'minor planets' or *asteroids* revolving in the space between Mars and Jupiter. A large part of interplanetary space is also traversed by solitary eccentric travellers, of peculiar structure and composition, generally known as *comets*. Their gradual disintegration (very rapid on a cosmic timescale) leaves behind them swarms of *meteors* which may be seen as the 'shooting stars' of our night sky if they happen to be on a collision course with the Earth, in whose atmosphere their long cosmic journey ends. Near the Sun, and in the plane of planetary orbits, the interplanetary space contains a certain amount of cosmic dust of so fine a texture (1–10 micrometres in size) that a microscope would be necessary to see it. This dust, which pollutes interplanetary space, is probably being produced by condensation in close proximity to the Sun; the scattering of sunlight upon it gives rise to the feebly luminous *zodiacal cloud* around the Sun and, in the anti-solar direction, to the *gegenschein*.

A brief account of these different constituents encountered in the solar system discloses that this entire system can be divided into two distinct parts: the realm of the *inner* planets and other bodies in the space between the Sun and the asteroidal belt, among which the largest and most massive astronomical body happens to be the Earth (hence the name of the *terrestrial* planets often given to bodies of this group); and the domain of the *major* planets, lorded over by Jupiter, which surrounds that of the terrestrial planets at a considerable distance from the Sun. All planets of both groups revolve around the Sun in the same direction and in elliptical orbits which deviate the less from circles, the larger their masses. These orbits cluster around a preferential plane, inclined by about 7° to the solar equator, with a dispersion diminishing again with the mass. However, the similarity of both groups is restricted only to these kinematic characteristics; their physical and chemical compositions could not be more different, the reasons probably being connected with their different locations in the system.

First let us consider the masses of the two groups of planetary bodies. The mass of Jupiter, the most massive planet of the solar system, amounts to 0·000 955 or approximately one-thousandth of that of the Sun and together with that of all the other planets of this group (Saturn, Uranus and Neptune) it adds up to about 0·13% of the solar mass. In spite of this disparity, the chemical composition of these planets is still essentially of the solar type, consisting as it does mainly of hydrogen and helium (in different states), with appreciable admixture of the elements of the carbon–nitrogen–oxygen group in Uranus and Neptune. In contrast, the masses of all the terrestrial planets (including the Moon) add up to no more than twice the mass of the Earth, or 0·5% of that of Jupiter, this latter disparity being not much smaller than that between Jupiter and the Sun! And even greater is the contrast between the chemical composition of the two groups of planetary bodies: while the major planets consist predominantly of hydrogen and helium, the terrestrial planets consist largely of oxygen, silicon and iron. Consequently, their mean densities lie between $5.52\,\mathrm{g\,cm^{-3}}$ for the Earth and $3.34\,\mathrm{g\,cm^{-3}}$ for the Moon, in contrast with densities between 0.69 and $1.56\,\mathrm{g\,cm^{-3}}$ for the major planets.

So conspicuous a disparity in composition between these planets is not a simple consequence of their masses. The ratio of 0.046 which the mass of Uranus bears to that of Jupiter is actually smaller than the ratio of 0.068 which the mass of the Earth bears to that of Uranus, and yet the mean density, increasing only from $1.34\,\mathrm{g\,cm^{-3}}$ for Jupiter to $1.56\,\mathrm{g\,cm^{-3}}$ for Uranus, jumps to $5.52\,\mathrm{g\,cm^{-3}}$ for the Earth. A different location in the solar system—impeding an escape of more volatile elements from Uranus or Neptune because of the low temperatures prevalent further away from the Sun—may constitute part of the answer, but probably not all of it.

How large is the solar system? If the Sun were alone in the Universe, its gravitational confines would extend to infinity. However, in the spiral arm of our Galaxy of which it is a part, it has to share its sphere of influence with the baronial estates of neighbouring stars. At the present stage of evolution of our Galaxy, particles of any mass can be regarded as at least fellow travellers, if not regular members, of the solar-system club if their distance from the Sun is of the order of 100 000 astronomical units. On occasion the Sun may exchange some fellow travellers (especially comets!) with nearby stars to become members of their own gravitational domains. But at distances smaller than 10 000 AU from the Sun—still very large in comparison with the space populated by the known planets—any inhabitants must, in general, be regarded as permanent members of the solar system.

Emigration from even these inner precincts of our system is also possible under exceptional circumstances: the passports of comets require validation only by the Jupiter or Saturn gravitational perturbations to move out where they please, but bodies of planetary size would find escape much more difficult. Nevertheless, evidence will be given in Chapter 6 that these exceptional circumstances may have actually arisen in the past. What we observe today may be only the remains of what could once have been a more complicated system, although this is still a matter of conjecture.

But whatever the case may have been, the system now known to us, whose permanent members are not much more than some 50 AU distant from the Sun, represents by volume only a very small part of the solar gravitational bailiwick. And even of that only a part will be of interest to us in this book: namely, the realm of the *terrestrial planets* located in the innermost precincts of the system and comprising planets between Mercury and Mars. By mass, the bodies we encounter in the immediate neighbourhood of the Sun constitute only a very small fraction of not only the mass of the solar system as a whole, but also of that of its planetary family, for the giant planets (Jupiter, Saturn, Uranus and Neptune) outweigh the combined mass of the terrestrial planets (Mercury, Venus, Earth and Mars) by more than 225 times.

On the other hand, by their chemical composition as well as their location in the solar system, the terrestrial planets constitute a fairly homogeneous family whose properties set it apart from other classes of celestial bodies. And, last but not least, in one other respect do these planets deserve the attention we propose to devote to them in this volume: because of their proximity to us, they have already become targets of successful and extensive exploration *in situ* by means of spacecraft. The results obtained with the aid of the space probes of the past 20 years read indeed like a fairy tale, and their import still continues to make us gasp in awe. In Chapters 2–8 we shall unfold the principal parts of this fairy tale for the non-technical reader, with the hope of recapturing by word and image at least part of the excitement which followed the acquisition of these recent revelations.

Limitations of space prevent us from developing the growth of our present knowledge more fully in historical perspective. Most of this knowledge has been acquired only since the methods of traditional astronomical approach at a distance were supplemented by more modern methods of radio, radar, and space astronomy based on the use of spacecraft. Indeed, the present state of our knowledge—as reflected in this book—could have been reached only by a continuing, concerted, interdisciplinary effort with the active participation of investigators from many different branches of science. It was the radio-physicists and chemists, geologists and engineers, as well as astronauts operating in the field who, together with the astronomers at their telescopes, have brought about the advances which we shall now proceed to describe.

2 Exploration by Spacecraft

Astronomy constitutes one of the oldest manifestations of awakening human intelligence, but has always been debarred from the status of a genuine experimental science by the remoteness of the objects of its study. With the exception of meteors—those freaks of cosmic matter intercepted by the Earth on its journey through space which have (occasionally) found their way into our laboratories—the properties of all celestial bodies could be investigated only at a distance, usually from the effects of attraction extended by their masses and from the ciphered messages of their light carried by nimble-footed photons across the intervening gaps of space.

The dramatic emergence of long-range spacecraft capable of defying terrestrial gravity and carrying not only telescopes and other scientific instruments, but also men, beyond the confines of our atmosphere completely changed this time-honoured picture. The 'beeps' of the first sputnik that began to circle the world on 4 October 1957 meant even more for the future of mankind than did the 'shots at Lexington' whose echo, in Emerson's words, went around the world in 1776.

The first extraterrestrial target of this new space age, which the parallel advances of rocket propulsion, long-range communication and computer control have so suddenly thrust upon us, was our nearest celestial neighbour—the Moon. In the wake of the brave attempts of the American rockets Pioneers 1 and 3 (launched in October and December 1958), it was the Russian Luna 1 (launched on 2 January 1959) which first 'made' it all the way and passed by the Moon at a distance of less than 6000 km (i.e. 2·4 times the Moon's radius), eventually to become the first artificial 'asteroid' of the Sun launched by man. Moreover, its direct successor, the Russian Luna 2 (launched eight months later on 11 September 1959) made history by scoring the first hard impact on the Moon on 13 September 1959—a memorable date in the history of mankind—after a flight lasting $63\frac{1}{2}$ hours. Less than a month later, Luna 3 (launched on 4 October in a highly eccentric Earth orbit) actually circumnavigated the Moon and gave us our first glimpse of its elusive far side.

The years since 1959 have been full of accomplishment and have provided the foundations on which lunar science rests at the present time. More complete basic data on the lunar spacecraft and their missions are listed in tables 1–3. In what follows we shall describe more fully the aims and individual accomplishments of these missions.

As far as the Moon is concerned, the years of 1960–1965 were essentially those of hard-landing and fly-by missions of American as well as Russian origin, which not only unveiled the far side of the Moon, but also provided, in the last seconds of their flight, glimpses of the structure of the lunar surface on a metre scale and at a resolution exceeding that attainable from the Earth by a factor of several hundred. The virtual absence of any lunar magnetic field, as well as the value of the Earth–Moon mass-ratio to a precision better than a hundred times our previous estimate of this important quantity, are among the more important determinations made by this class of spacecraft. The data so obtained will be reported in Chapter 3, while table 2 given here contains some of the 'astronomical' data concerned with their landings. Only those spacecraft which provided scientific information are included; the local values of lunar radii determined from the exact timings of the respective impacts are of particular interest.

With the impact of Ranger 9 in the crater Alphonsus on 24 March 1965 the contributions of the lunar hard-landers came to an abrupt end, for the next three years witnessed the rise of two other families of mooncraft whose contributions were to dominate the field till 1968—the lunar soft-landers (plate 2) and the lunar orbiters (plate 3). The former preceded the latter by some months, though not without mishap. Five attempts at soft landings (Lunas 4–8) were made by the Russians before the success of Luna 9 on 3 February 1966, a success repeated by Luna 13 shortly before the end of the same year. The first American soft-lander, Surveyor 1, was totally successful. Surveyor 2, launched in September 1966, failed en route because of equipment malfunctioning. Surveyor 3, launched in April 1967, was successful, though Surveyor 4 failed through loss of communications only $2^1/_2$ minutes before landing. However, the fifth soft-landed safely in September 1967; the sixth in November; and the seventh (the last one) in January 1968.

With the soft-landing Surveyors and Lunas whole scientific laboratories capable of performing a variety of experiments *in situ* reached the lunar surface. On-board television cameras provided us with tens of thousands of frames, ranging from close-up shots of the surface attaining a resolution of the order of 1 mm in the immediate proximity of the spacecraft's landing pads (cf plate 4), to views of the solar corona at sunset (plate 5) or of the total eclipses of the Sun by the Earth witnessed on the Moon by Surveyor 3 on 24 April 1967 and by Surveyor 5 on 18 October of the same year. Other experiments aboard included samplers of the mechanical properties (compressibility, graininess) of the lunar soil and, on Surveyors 5–7, chemical samplers capable of analysing the atomic composition of the lunar surface by observing the scattering of α-particles from the lunar ground. It was this experiment—successful on all three spacecraft—which, in the hands of A Turkevich and his associates, suggested for the first time that the rocks

Table 1. List of lunar spacecraft (1959 – 1976)

Name of spacecraft	Origin	Date of launch §	Weight (kg)†
(a) Fly-by			
Luna 1	USSR	2 January 1959	362
Luna 3	USSR	4 October 1959	435
Ranger 3	USA	26 January 1962	330
Ranger 5	USA	18 October 1962	342
Luna 4	USSR	2 April 1963	1422
Luna 6	USSR	8 June 1965	1442
Zond 3	USSR	18 July 1965	960
(b) Hard-landing			
Luna 2	USSR	11 September 1959	390
Ranger 4	USA	23 April 1962	331
Ranger 6	USA	30 January 1964	365
Ranger 7	USA	28 July 1964	366
Ranger 8	USA	17 February 1965	367
Ranger 9	USA	21 March 1965	367
Luna 5	USSR	9 May 1965	1476
Luna 7	USSR	4 October 1965	1506
Luna 8	USSR	3 December 1965	1552
Surveyor 2	USA	20 September 1966	990
Surveyor 4	USA	14 July 1967	1038
(c) Soft-landing			
Luna 9	USSR	31 January 1966	1583 (100)
Surveyor 1	USA	30 May 1966	990 (292)
Luna 13	USSR	21 December 1966	1580 (100?)
Surveyor 3	USA	17 April 1967	1040 (302)
Surveyor 5	USA	8 September 1967	1006 (303)
Surveyor 6	USA	7 November 1967	1008 (303)
Surveyor 7	USA	7 January 1968	1010 (305)
Luna 17	USSR	10 November 1970	? (755)
Luna 21	USSR	8 January 1973	? (840)
(d) Orbiting			
Luna 10	USSR	31 March 1966	245
Orbiter 1	USA	10 August 1966	387
Luna 11	USSR	24 August 1966	1640
Luna 12	USSR	22 October 1966	?
Orbiter 2	USA	6 November 1966	390
Orbiter 3	USA	5 February 1967	385
Orbiter 4	USA	4 May 1967	390
Explorer 35	USA	19 July 1967	104
Orbiter 5	USA	1 August 1967	390
Luna 14	USSR	7 April 1968	?
Luna 15	USSR	13 July 1969	?
Luna 18	USSR	2 September 1971	?
Luna 19	USSR	28 September 1971	?
Luna 22	USSR	29 May 1974	4000
Luna 23	USSR	28 October 1974	4000

Table 1—continued

Name of spacecraft	Origin	Date of launch§	Weight (kg)†
(e) Re-entrant			
Zond 5	USSR	15–21 September 1968	Unmanned
Zond 6	USSR	10–17 November 1968	Unmanned
Apollo 8	USA	21–27 December 1968	Manned
Apollo 10	USA	18–26 May 1969	Manned
Apollo 11	USA	16–24 July 1969	Manned‡
Zond 7	USSR	9–15 August 1969	Unmanned
Apollo 12	USA	14–24 November 1969	Manned‡
Apollo 13	USA	11–17 April 1970	Manned
Luna 16	USSR	12–24 September 1970	Unmanned
Apollo 14	USA	31 January–9 February 1971	Manned‡
Apollo 15	USA	26 July–7 August 1971	Manned‡
Luna 20	USSR	14–25 February 1972	Unmanned
Apollo 16	USA	16–27 April 1972	Manned‡
Apollo 17	USA	6–19 December 1972	Manned‡
Luna 24	USSR	9–22 August 1976	Unmanned

†The weights in parentheses given for the soft-landing spacecraft refer to those of the instrumented packages actually deposited on the lunar surface.
‡Manned landings on the surface.
§Range of dates for re-entrant spacecraft indicates duration of mission.

Table 2. Place and time of unmanned spacecraft landing on the Moon†

Spacecraft	Place of impact		Date and time of impact‡	
	Longitude	Latitude		
(a) Hard-landing				
Luna 2	0°0	29°1N	13 September 1959	$22^h02^m24^s$§
Ranger 6	21°52E	9°33N	2 February 1964	$9^h24^m33^s.1$
Ranger 7	20°58W	10°35S	31 July 1964	$13^h25^m49^s$
Ranger 8	24°65E	2°67N	20 February 1965	$9^h57^m36^s.8$
Ranger 9	2°37W	12°83S	24 March 1965	$14^h08^m20^s$
(b) Soft-landing				
Luna 9	64°37W	7°08N	3 February 1966	$18^h44^m52^s$
Surveyor 1	43°21W	2°45S	2 June 1966	$6^h17^m37^s$
Luna 13	62°05W	18°87N	24 December 1966	18^h01^m
Surveyor 3	23°34W	2°94S	20 April 1967	$0^h04^m53^s$
Surveyor 5	23°18E	1°41N	11 September 1967	$0^h46^m44^s.3$
Surveyor 6	1°37W	0°46N	10 November 1967	$1^h01^m5^s.5$
Surveyor 7	11°41W	41°01S	10 January 1968	$1^h05^m30^s$
Luna 17	35°0W	38°28N	17 November 1970	3^h47^m
Luna 21	30°38E	25°51N	15 January 1973	23^h35^m

†Only those spacecraft which furnished lunar scientific information are included.
‡Universal Time as observed on the Earth (not corrected for transit time of the signals).
§The last stage of the carrier rocket of Luna 2 (1121 kg in weight) impacted on the Moon 30 minutes later.

Table 3. Kinematic characteristics of the artificial lunar satellites (1966 –1974)

Spacecraft	Period	Inclination[†]	Altitude (km)[‡] Periselenium	Aposelenium	Injection into lunar orbit	End of mission	Number of days in orbit	Total number of revolutions
Luna 10	$178^m.3$	$71°.9$	349	1017	3 April 1966	30 May 1966	67	—
Orbiter 1	$208^m.6^s$	$12°$	56	1853	14 August 1966	29 October 1966	76	547
Luna 11	178^m	$27°$	159	1200	18 August 1966	1 October 1966	34	—
Luna 12	205^m	$4°$	105	1740	26 October 1966	—	—	—
Orbiter 2	$208^m.4^s$	$11°.9$	49.7	1853	10 November 1966	11 October 1967	335	2289
Orbiter 3	$208^m.6^s$	$20°.9$	54.9	1847	8 February 1967	9 October 1967	243	1843
Orbiter 4	721^m	$85°.5$	2706	6114	8 May 1967	6 October 1967	70	225
Explorer 35	684^m	$11°.2$	763	7670	19 July 1967	in orbit	—	—
Orbiter 5	$510^m.5^s$	$85°$	195	6029	5 August 1967	—	179	—
Orbiter 5	$503^m.5^s$	$84°.6$	100	6066	5 August 1967	—	179	—
Orbiter 5	$191^m.3^s$	$84°.8$	99	1499	5 August 1967	29 January 1968	179	1201
Luna 14	160^m	$42°$	159	871	10 April 1968	—	—	—
Luna 15	150^m	$126°$	132	287	16 July 1969	21 July 1969	5	52
Luna 18	119^m	$35°$	96	101	7 September 1971	11 September 1971	5	54
Luna 19	$121^m.8^s$	$40°.6$	139	—	3 October 1971	30 September 1972	368	4350
Luna 22	192^m	$19°.6$	171	1437	2 June 1974	11 November 1974	162	1778
Luna 23	—	—	—	—	31 October 1974	6 November 1974	7	96

[†]Inclination to lunar equator.
[‡]Altitude above the mean lunar sphere of radius 1738 km.
[§]Date of loss of radio contact.

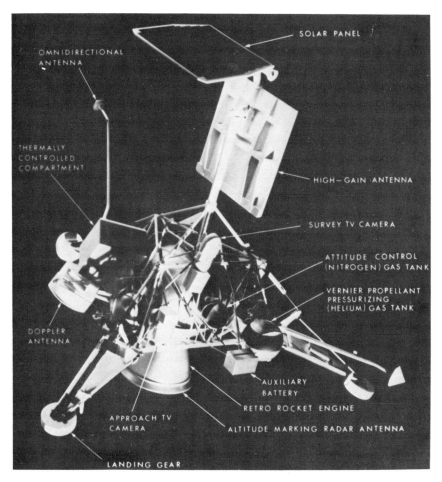

Plate 2. A schematic view of the American soft-landing Surveyors of 1966–1967 showing the location of different instrumental packages aboard the spacecraft. *Photograph by courtesy of NASA and JPL.*

on the lunar surface are chemically differentiated (a suggestion fully confirmed by a laboratory analysis of rocks brought back by the subsequent Apollo missions).

With the exception of one 'hop' performed by Surveyor 6 on 17 November 1967 by use of its own retro-rockets, none of the spacecraft which soft-landed on the Moon between 1966 and 1968 was capable of locomotion. The possibility of sending a roving vehicle to the Moon was briefly considered in the United States in the early 1960s under the name of the short-lived Project Prospector, but the project was dropped so as not to compete with the requirements of the Surveyors. In fact, it was not until November 1970 that a roving vehicle reached the Moon in the form of the first of the Russian Lunokhods (Luna 17), an event repeated in January

1973 by another Lunokhod (Luna 21) which landed in a different part of the lunar surface.

While the soft-landers of 1966–1968 and the rovers of 1970–1973 carried multipurpose laboratories to the Moon, the tasks of the lunar orbiters have since 1966 been substantially simpler (table 3): they

Plate 3. A view of the American Orbiter spacecraft of 1966–1967 in the laboratories of the Aerospace Division of the Boeing Company in Seattle, with its antennae and solar panels unfolded. The optics of the Orbiter cameras are exposed to view in the front. *Photograph by courtesy of NASA and Boeing.*

were to provide high-resolution photographic coverage of the lunar surface as a prerequisite for a choice of landing sites for the manned missions of 1969–1972. This was successfully accomplished by the American orbiters (see plate 3), whose combined photographic output (scanned aboard the spacecraft and subsequently televised to the Earth for reconstitution) provided us with almost as good a permanent record of the lunar surface as we then possessed for the Earth. Several examples of their work will be reproduced in the next chapter (see plates 12, 15 and 17).

Apart from their purely photographic role, the orbiters—which described more than 6000 revolutions around the Moon between 1966 and 1968—also provided by their motions sensitive indications of the lunar gravitational field. Accurate radio tracking of the motions of the American orbiters and its analysis led, in particular, to P Muller and W L Sjogren's discovery in 1968 of the existence of localized concentrations of mass ('mascons') at a shallow depth below the lunar surface, thus providing an important tool for the subsequent examination of the interior of our satellite. An especial place among lunar orbiters (the kinematic characteristics of

Plate 4. An imprint of one of the footpads of Surveyor 3 in the lunar soil, caused by recoil on landing on 20 April 1967, and photographed *in situ* by the astronauts of the Apollo 12 mission three years later. *Photograph by courtesy of NASA and Hasselblad.*

which are summarized in table 3) is occupied by the American Explorer 35, launched in July 1967, whose mission was not photographic, but magnetic. It was to measure the possible lunar magnetic field and its interaction with the 'solar wind' (the results obtained are referred to later in the text).

A retrospective look at all work carried out since 1959 (summarized in tables 1–3) discloses that the exploration of the Moon by spacecraft has so far been the exclusive domain of the superpowers—the USA and the USSR. No other nation has so far been able to add any direct contribution and is unlikely to do so for many years to come. Furthermore, it can be noted that in all feats accomplished up to 1966 the major achievements belonged to the Russians: theirs was the first lunar fly-by (Luna 1 in January 1959);

Exploration by Spacecraft 21

Plate 5. Sunset on the Moon: a view of the solar corona and its streamers as televised by Surveyor 1 on 13 June 1966. *Photograph by courtesy of NASA and JPL.*

the first hard impact (Luna 2 in September 1959) or circumnavigation (Luna 3 in October of the same year); the first soft landing (Luna 9 in February 1966) as well as the first lunar orbiter (Luna 10 in March of the same year); the first re-entrant spacecraft (September 1968); and the first lunar rover (November 1970).

However, the extent of the Russian lead over parallel American achievements was progressively diminishing. While it took more than two years after the Russian feat for an American spacecraft to fly past the Moon (Ranger 3 in January 1962) or to score a direct hit (Ranger 4 in April of the same year on the Moon's far side), the time of the Russian lead for soft landings or injection of spacecraft into circumlunar orbits had already been reduced to a few months. Moreover—and this is more important—the American contributions, when they came, were on so massive a scale as to provide most of the evidence we now possess. This disparity should certainly not diminish the historical significance and pioneering nature of much of the work done by the USSR in space. Its achievements paved the way and, once the feasibility became a proven fact, its follow-on was no doubt encouraged by the knowledge that it could be done. Without the Russian sputniks of 1957 there would probably be no National Aeronautics and Space Administration in the United States; and without the Russian moon probes of 1959, in 1961 President Kennedy would have scarcely proclaimed the Moon a target to be reached by American astronauts before the end of that decade.

But proclaim it he did, and the task was brilliantly accomplished between 1969 and 1972 by Project Apollo, which has already gone down in history as one of the greatest achievements of our age and one which is likely to be remembered by posterity long after all the more ephemeral events of the recent past have been consigned to oblivion. Because of the need to provide a life-support system for the three participating astronauts, a manned return trip to the Moon constituted a much more difficult task than landing an unmanned space probe. To transport these men to the Moon, the Apollo planners had to design and build a series of giant rockets (Saturn C-5) at NASA's Marshall Space Flight Center in Huntsville, Alabama, under the direction of that inveterate enthusiast Wernher von Braun (1912–1977), who pursued this goal with a single-minded devotion throughout his whole career in different parts of the world and sometimes at risk of his life.

This great *Columbiad* of the twentieth century now belongs to the past, but perhaps a brief description of its carrier rocket may be of interest here. Plate 6 shows the rocket on the launching platform at Cape Kennedy, Florida, ready to take off on a million-kilometre round-trip to the Moon. Its first stage, 13 m in diameter, towered 51 m above the ground, while the second and third stages, approximately 24 m in height and 6·5 m in diameter, completed the structure almost 100 m high, about as tall as the spires of the largest Gothic cathedrals erected in the Middle Ages. The total weight of this spacecraft was about 3000 tons at take-off and equivalent to a combined weight of 25 B-55 bombers or, for those who prefer to think in naval terms, of one light cruiser.

To sail a cruiser at 30–40 knots is one thing, but to lift it vertically and accelerate it in a few minutes to a velocity of $7·91$ km s^{-1} (necessary to place it in a circumterrestrial orbit preparatory to travel to the Moon) is obviously another; to accomplish this vastly more powerful engines are necessary. The Saturn rocket was lifted from the ground by a cluster of five jet engines whose combination supplied a thrust of $7^{1}/_{2}$ million pounds, equivalent to almost 150 million standard horse power. Needless to say, to deliver this power by chemical means required a prodigious amount of fuel: while the engines of the first stage were burning for about $2^{1}/_{2}$ minutes, the Saturn rocket consumed some 15 tons per second of a mixture of two parts of kerosene and one part of liquid oxygen. Its total of 2250 tons of fuel used up during this time was large enough to drive an average-size family car beyond the limits of the solar system! After the first stage had lifted the Saturn

Exploration by Spacecraft 23

Plate 6. Anchors away! A Saturn C-5 rocket launching a mission on a million-kilometre round-trip to the Moon. *Photograph by courtesy of NASA.*

assembly to an altitude of about 100 km (at which atmospheric resistance becomes negligible), the second stage, equipped with five engines burning a hydrogen–oxygen mixture, injected a 120 ton payload into orbit around the Earth at an altitude of 200 km. The third stage, powered by one engine of this latter type, then accelerated the remaining 45 ton payload to the Earth's escape velocity of $11\cdot2$ km s^{-1} and sent it on its way to the Moon.

The journey to the Moon, during which the spacecraft remained

Table 4. Place and time of re-entrant spacecraft landing on the Moon

Mooncraft	Place of landing Region	Longitude	Latitude	Date of landing	Time of landing
Apollo 11	Mare Tranquillitatis	23°29'24"E	0°40'12"N	20 July 1969	$6^h 14^m$
Apollo 12	Oceanus Procellarum	23°20'23"W	2°27'0"S	18 November 1969	$20^h 52^m$
Luna 16	Mare Foecunditatis	56°18'E	0°41'S	20 September 1970	$5^h 18^m$
Apollo 14	Fra Mauro	17°27'55"W	3°40'24"S	5 February 1971	$8^h 37^m 10^s$
Apollo 15	Mare Imbrium	3°39'10"E	26°6'4"N	30 July 1971	$22^h 16^m 29^s$
Luna 20	Mare Foecunditatis	56°30'E	3°34'N	21 February 1972	$19^h 19^m$
Apollo 16	Descartes	15°30'47"E	8°59'34"S	21 April 1972	$2^h 23^m 36^s$
Apollo 17	Taurus–Littrow	30°45'26"E	20°9'41"N	11 December 1972	$19^h 54^m 57^s$
Luna 24	Mare Crisium	62°12'E	12°15'N	18 August 1976	$7^h 36^m$

Table 5. Manned flights to the Moon (1968–1972)

Mission	Date of start of mission	Participating astronauts[†]	Duration of entire mission	Duration of stay on lunar surface
Apollo 8	21 December 1968	F Borman, J A Lovell, W A Anders	$147^h 0^m$	—
Apollo 10	18 May 1969	T P Stafford, J W Young, E A Cernan	$192^h 03^m$	—
Apollo 11	16 July 1969	N A Armstrong, M Collins, E E Aldrin	$195^h 18^m$	$21^h 36^m$
Apollo 12	14 November 1969	Ch Conrad, R F Gordon, A L Bean	$244^h 36^m$	$31^h 30^m$
Apollo 13	11 April 1970	J A Lovell, J L Swigert, F W Haise	$142^h 55^m$	—
Apollo 14	31 January 1971	A B Shepard, S A Roosa, E D Mitchell	$216^h 02^m$	$33^h 31^m$
Apollo 15	26 July 1971	D R Scott, A M Worden, J B Irwin	$295^h 12^m$	$66^h 56^m$
Apollo 16	16 April 1972	J W Young, T K Mattingly, Ch M Duke	$265^h 51^m$	$71^h 02^m$
Apollo 17	6 December 1972	E A Cernan, R E Evans, H H Schmitt	$301^h 52^m$	$74^h 59^m$

[†]The first name is that of the commander of the respective mission and the second that of the command module pilot (i.e. orbiting astronaut).

Plate 7. Apollo 11 astronaut Edwin E Aldrin deploying the passive seismic experiment package on the surface of the Moon on 20 July 1969. The cylinder in the middle of the package houses the actual seismometer which is flanked on both sides by arrays of solar cells to provide electric power for operation and transmission. *Photograph by courtesy of NASA.*

largely in free flight, lasted about 65–70 hours. When it approached the Moon it was decelerated by retro-rockets to enable it to be captured by the Moon in a closed orbit, in which it remained until the time came for the last scene of this first act of lunar astronautics. Again by the action of rockets, the spacecraft separated into two parts: the command and service module, which remained in orbit with one astronaut aboard; while the other, the excursion module carrying two astronauts, descended to the lunar surface to perform its appointed tasks. After its part of the mission was accomplished, the excursion module (see colour plate 4) was lifted from its platform by rockets to regain an orbital altitude at which it could rendezvous with the command module (colour plate 1(*b*)). The three astronauts, then happily

Plate 8. The lunar retro-reflector of the Apollo 14 mission with its square array of 100 cube corners for returning laser flashes back to the Earth. *Photograph by courtesy of NASA.*

reunited, accelerated the remaining torso of their spacecraft for the last time on its homeward journey back to the Earth.

This is not the place to give detailed profiles of the nine successive Apollo missions which took off for the Moon between 1968 and 1972 and returned home safely; instead, the salient facts are listed in tables 4 and 5. The humane aspects of these historical encounters with our satellite have been described in numerous books which have appeared in many languages since 1969 and need not be repeated here. In what follows we shall therefore confine our attention to the scientific equipment which was deployed by the Apollo missions on the Moon and which furnished much of the data reported in Chapter 3. The pride of place among these should probably go to the seismic experiments (see plate 7), both active and passive, which constituted an essential part of the Apollo 11–16 missions. Some of these were still operative in later years of the decade, such as the cube-corner reflectors (plate 8) for the return of laser pulses beamed on them from the Earth.

As is well known, a beam of light incident on the corner of a cube is reflected successively from its three faces and returned in a direction exactly parallel with that of the incident beam. The only kind of light sent out from the Earth that has any chance of returning a measurable reflection is short pulses of radiation characterized by coherence in phase (laser). The parallelism of coherent beams largely ensures their freedom from attenuation with distance and will also ensure that a laser pulse reflected from the Moon will return to the vicinity of its orgin on the Earth. An accurate determination of the time-lag between the outgoing pulse and its returning 'echo' (which propagates through space at the speed of light) permits us to track the motion of the Moon with an accuracy unthinkable previously. The cube corners which made this possible were deposited on the Moon by the Apollo 11 and 14 missions, as well as by the unmanned Russian Lunokhods.

In addition to the instruments for seismic work and laser ranging, the Apollo Lunar Surface Experimental Package (ALSEP, see plate 9), installed on the Moon by the Apollo 12 and 14–17 missions, included magnetometers to measure local magnetic fields, a cold-cathode ion gauge, and a suprathermal ion detector for studies of the particulate contents of the lunar environment. Other devices were designed to study the heat flow through sub-surface layers (Apollo 15 and 17), the electrical properties of the lunar surface (Apollo 17) and the mechanics of lunar soil (Apollo 14–17).

Needless to say, ample attention was given by each mission to field geology and photography, tasks in which the astronauts of the last three missions (Apollo 15–17) were greatly assisted by the availability of lunar-roving vehicles (colour plate 2), which enabled them to extend their range of exploration far beyond the radius of action of the first three manned missions. One of the most important tasks of the astronauts on the Moon was to collect representative samples of lunar rocks for subsequent analysis in terrestrial laboratories. This they did with great success: the 382 kilograms of lunar rocks brought back to the Earth by the Apollo 11–17 missions became by far the best studied material in the history of petrology and geochemistry, with results which amply justified the effort and on which we shall have more to say in the next chapter.

Our present account of the research programmes carried out by the six manned Apollo missions has so far been limited to activities on the lunar surface. But only two astronauts of each mission actually descended to the ground; the third remained in the command module in circumlunar orbit. On the first two missions, and partly also on the third, this astronaut in orbit was the 'forgotten man' of the project as far as lunar science was concerned. It was not until the Apollo 15 mission that the orbiting astronaut

Plate 9. The Apollo Lunar Science Experimental Package (ALSEP) of the Apollo 14 mission *in situ* on the Moon near the crater Fra Mauro. *Photograph by courtesy of NASA.*

also became an important participant in the lunar exploration. His task was not only to secure photographs of unparalleled quality and resolution of the lunar surface overflown by the command module, but also to operate an array of other experiments which would contribute to the scientific significance of each mission.

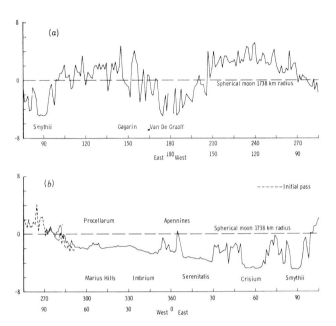

Figure 1. Hypsometric profiles of the lunar globe established by the laser altimeter of the Apollo 15 mission, exhibiting the deviations of the actual shape of the Moon's near (b) and far (a) side from a sphere of 1738 km radius along the track overflown by this spacecraft. Note a general depression of the near-side hemisphere below the mean moon-level as well as a systematic incline of the levels of the maria—from Imbrium to Smythii—on the near side.

One of the most important instruments of the Apollo 15 and 16 missions in orbit around the Moon was a laser altimeter, a device which measured the vertical deformations of the lunar surface along tracks on the ground overflown by the spacecraft (see figures 1 and 2). Other instruments were an x-ray fluorescence spectrometer to study the flux of x-rays produced on the lunar surface by the impact of the solar wind (and, through it, its large-scale composition); x-ray and γ-ray spectrometers to measure the natural radioactivity of the lunar surface; and a mass spectrometer to establish the nature of the atomic constituents of the lunar exosphere. The command modules of the Apollo 15 and 16 missions did not orbit the Moon alone, but were accompanied by sub-satellites with their own scientific

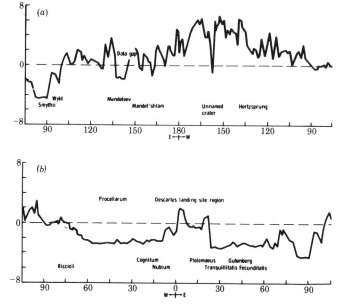

Figure 2. Hypsometric profiles of the lunar globe established by the laser altimeter of the Apollo 16 mission, exhibiting the extent of the deviations of the actual shape of the Moon's near (b) and far (a) side from a sphere of 1738 km radius along the track overflown by this spacecraft.

instrumentation concerned with the motion of the pair, with the shadow cast by the Moon in the solar wind, and with the measurement of the lunar magnetic field.

It should be added that, while the American astronauts visited the surface of the Moon and returned home in triumph no less than six times between 1969 and 1972, the Russians did not remain idle. Even before Apollo 8 had paid a brief orbital visit to the Moon on Christmas 1968, the unmanned spacecraft Zond 5 and 6 accomplished similar manoeuvres around the Moon in September and October of the same year. And it was only a technical difficulty which prevented Luna 15 of July 1969 from returning to the Earth before Apollo 11 with the first sample of lunar rocks collected automatically—a feat accomplished subsequently by the re-entrant Lunas 16, 20 and 24.

By their combined output, all the re-entrant spacecraft—both manned and unmanned—of the last decade have increased our knowledge of the Moon and its environment immeasurably. However, no comparison of the effectiveness of the manned versus unmanned tools of research can leave any doubt that, in the period between 1969 and 1972, man amply justified his place in the system by securing results which could not have been obtained by any automatic device known at that time. To give an example, the six Apollo missions which landed on the Moon brought back 1000 times as large a mass of lunar rocks as did the unmanned Lunas 16, 20 and 24. Moreover, while the Apollo samples came from localities which were judiciously chosen, the Lunas had to lift them from the ground on which they happened to land. Within the 'state of the art' of human technology between 1968 and 1972, man on the Moon was still a necessity for the rapid advances of lunar science which we experienced at that time. Whether or not this will also be true in the future remains, however, an open question.

Planetary Space Probes

As their name alone suggests, the lunar spacecraft listed in table 1 had one destination in common: the fly-by's of group (*a*) in the table represent near misses; all other craft either crash-landed on the Moon—the hard-landers of group (*b*) having done so on arrival and the orbiters of group (*d*) at a later time—or soft-landed (group (*c*)) to remain on the Moon for ever. The only exceptions were the re-entrant soft-landers of the Apollo group which returned to the Earth to effect another soft landing in the Pacific Ocean, and the Russian soft-landers of group (*e*) which returned to their launching places in Central Asia. Luna 3 was really an Earth satellite in a highly eccentric orbit, and eventually met its fiery end in our atmosphere; while Explorer 35, a distant orbiter, continues to keep its lonely vigil around the Moon in anticipation of future developments in cislunar space. With the exception of Luna 3, all fly-by's of group (*a*), after having passed by the Moon, went into *heliocentric* orbits. Luna 1 of 1959 was destined to become the first 'asteroid' manufactured by man.

Moreover, the same was true of a majority of the space probes sent out to reconnoitre the planets Mercury, Venus and Mars on missions summarized in table 6. The first target was Venus—our nearest planetary neighbour (see Chapter 7). The first successful fly-by was accomplished by the American Mariner 2 on 14 December 1962, while the first crash-landing on the surface of this planet was effected by the Russian Venera 3 on 1 March 1966. Although the latter failed to return any data, it became a harbinger of greater things to come, for in its wake came a series of subsequent Veneras (4–9) whose descent through the Cytherean atmosphere was decelerated by parachutes (see figure 3). During the descent these spacecraft transmitted data on pressure, temperature and composition of the atmosphere. The latest member of this series, Venera 9 of October 1975, provided the first televised views of the Cytherean surface (see plate 54).

It should also be mentioned here that, after paying a close call on Venus on 5 February 1974, the American Mariner 10 went on to effect repeated fly-by's past Mercury (on 29 March and 21 September 1974, followed by one on 16 March 1975) and for the first time gave us knowledge

Table 6. Planetary probes launched between 1961 and 1978

Name	Origin	Date of launch	Destination
Venera 1	USSR	12 February 1961	Venus (contact lost after 7·5 million km)
Mariner 2	USA	26 August 1962	Venus (fly-by on 14 December 1962)
Mars 1	USSR	1 November 1962	Mars (contact lost after 105 million km)
Zond 1	USSR	2 April 1964	Venus (contact lost en route)
Mariner 3	USA	5 November 1964	Mars (failed by shroud malfunction)
Mariner 4	USA	28 November 1964	Mars (fly-by on 14 July 1965)
Zond 2	USSR	30 November 1964	Mars (batteries failed on 5 May 1965)
Venera 2	USSR	12 November 1965	Venus (fly-by on 27 February 1966)
Venera 3	USSR	16 November 1965	Venus (crash-landing on 1 March 1966)
Venera 4	USSR	12 June 1967	Venus (parachute landing on 18 October 1967)
Mariner 5	USA	14 June 1967	Venus (fly-by on 19 October 1967)
Venera 5	USSR	5 January 1969	Venus (parachute landing on 16 May 1969)
Venera 6	USSR	10 January 1969	Venus (parachute landing on 17 May 1969)
Mariner 6	USA	24 February 1969	Mars (fly-by on 31 July 1969)
Mariner 7	USA	27 March 1969	Mars (fly-by on 5 August 1969)
Venera 7	USSR	17 August 1970	Venus (parachute landing on 15 December 1970)
Mars 2	USSR	19 May 1971	Mars (ejection of capsule on 27 November 1971)
Mars 3	USSR	28 May 1971	Mars (capsule soft landing on 2 December 1971)
Mariner 9	USA	30 May 1971	Mars (orbiter since 13 November 1971)
Pioneer 10	USA	3 March 1972	Jupiter (fly-by on 4 December 1973)
Venera 8	USSR	26 March 1972	Venus (landed on 22 July 1972)
Pioneer 11	USA	6 April 1973	Jupiter (fly-by on 3 December 1974); to encounter Saturn in September 1979
Mars 4	USSR	21 July 1973	Mars (orbiter since January 1974)
Mars 5	USSR	25 July 1973	Mars (orbiter since January 1974)
Mars 6	USSR	5 August 1973	Mars (landed on 12 March 1974)
Mars 7	USSR	9 August 1973	Mars (contact lost on 12 March 1974)
Mariner 10	USA	3 November 1973	Venus (fly-by on 5 February 1974) Mercury (fly-by's on 29 March and 21 September 1974; 16 March 1975)
Venera 9	USSR	8 June 1975	Venus (landed on 16 October 1975)
Venera 10	USSR	14 June 1975	Venus (landed on 23 October 1975)
Viking 1	USA	20 August 1975	Mars (landed on 20 July 1976)
Viking 2	USA	9 September 1975	Mars (landed on 3 September 1976)
Voyager 1	USA	20 August 1977	Jupiter and Saturn (en route)
Voyager 2	USA	5 September 1977	Jupiter and Saturn (en route)
Pioneer Venus 1	USA	20 May 1978	Venus (to arrive on 4 December 1978)
Pioneer Venus 2	USA	7 August 1978	Venus (to arrive on 9 December 1978)

Exploration by Spacecraft 33

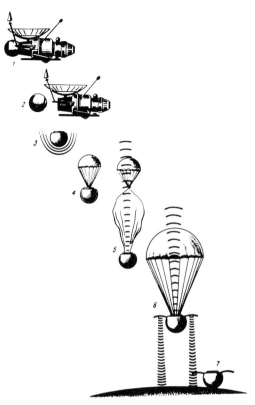

Figure 3. Schematic sequence of operations illustrating the descent through the atmosphere and soft landing of the Russian Veneras on the surface of Venus.

of the principal features of the surface of this planet, as well as of many other of its physical characteristics (see Chapter 4).

In contrast with Venus, whose space exploration gradually became almost a Russian domain—to which the Americans contributed with the successful fly-by's of Mariner 2 in 1962 and Mariner 10 in 1974—the planet Mars became largely an American preserve through the highly successful performances of Mariner 4 (fly-by on 14 July 1965) and Mariners 6 and 7 (fly-by's on 31 July and 5 August 1969). In November 1971 the American Mariner 9 became the first orbiter of another planet. It was joined in this function some days later by the Russian Mars 2 and 3 probes, and again by Mars 4 and 5 in January 1974. The Russians were also the first to crash-land a part of the Mars 2 probe on the surface of this planet, but their subsequent attempts to soft-land on the Martian surface with the Mars 3 or 6 probes by parachute descents were unsuccessful, as the landers failed to return any scientific data after touchdown. The accomplishment of this feat for the first time was reserved for the American Viking spacecraft of the summer of 1976 (see table 7). The results provided by them will be reviewed in Chapter 5; in what follows we merely wish to describe these spacecraft briefly.

By the calibre of their instrumentation, the Vikings were more elaborate probes than the lunar Surveyors of 10 years earlier, and in their integrity combined the functions of orbiters as well as soft-landers (see figure 4). Their total weight with fuel was not far from 5000 kilograms in terrestrial gravity and, because of this weight, the Titan–Centaur launcher

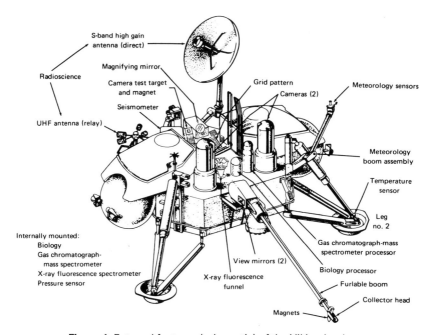

Figure 4. External features (schematic) of the Viking lander.

could not accelerate them to leave the Earth with the velocity of the earlier and more nimble-footed Mariners. As a result of their launch on 20 August and 9 September 1975, they had to spend almost 10 months en route before reaching the location of their celestial target. But when we stop to consider that their route was more than 460 million kilometres long and their time of flight corresponded to a daily progress of more than 1·5 million kilometres towards their goal, then this is not a bad record at all! In point of fact, both Vikings 'made it' on schedule almost within minutes. The first one was intended to land on Mars on 4 July 1976—the two-hundredth anniversary of the birth of the United States. However, fate willed otherwise; and the reasons for the delay became apparent soon after the injection of the system into a Martian orbit, when the television eyes of the orbiter began to scrutinize the prospective landing place (pre-selected on the basis of evidence supplied previously by Mariner 9).

At a closer look this place turned out to be so rough as to make a soft landing extremely hazardous. A more favourable spot was frantically searched for in the light of new evidence supplied by the Viking 1 orbiter. It was not until 20 July—more than two weeks after the date originally planned—that, on command from the Earth (a command which had to travel for 19 minutes through space to reach its destination), the lander part of the spacecraft decoupled from the orbiter and commenced its descent. It was decelerated first by parachutes and later by retro-rockets to bring it within 10 minutes into soft contact with the Martian surface. A sequence of the operations necessary to bring this about is shown diagrammatically in figure 5 and the spacecraft itself (*in situ*) in colour plate 6(*a*). The reader may, in

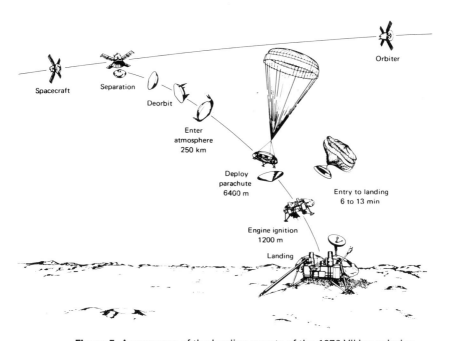

Figure 5. A sequence of the landing events of the 1976 Viking mission.

particular, note that at little more than 6 km above the surface the lander was still descending at a velocity of 900 km h^{-1}, a speed which over the next 5 km was reduced by parachute to 230 km h^{-1} (i.e. from the speed of a jet plane to that of a piston-propelled one of older vintage). Thereafter retro-rockets slowed the descending spacecraft further to a mere 2·5 m s^{-1} (corresponding to 9 km h^{-1}) at touchdown, a speed attained on Earth by free fall from an altitude of little more than one foot.

In addition to television cameras which could scan the Martian landscape from the horizon (about 3 km distant) to the immediate proximity of the spacecraft, the instrumentation aboard the Vikings included a seismometer (the only component which failed to function on Viking 1 but performed satisfactorily on Viking 2), an x-ray fluorescence spectrometer for analysis of the atomic composition of the Martian ground, and a gas chromatograph–mass spectrometer for the determination of its molecular composition (including a search for organic life). Further components of the equipment of these spacecraft contained assemblies for meteorological studies (i.e. measurements of temperature, as well as of wind direction and velocity), and those to study the mechanical properties of the soil, in addition, of course, to the radio links for receipt and transmission of information. The whole system was powered by a small nuclear power plant aboard the lander, delivering only about 65 watts of electrical energy but functioning day and night.

The Viking orbiters described highly eccentric orbits synchronous with the rotation of the planet (for their characteristics see table 8) and, in contrast with the landers, they carried within their payloads of 2325 kg elaborate optical systems for the imaging of the Martian ground at different resolutions, as well as infrared sub-systems for the thermal mapping of the surface and the detection of water in the atmosphere (for examples of the orbiters' work see colour plates 6(b) and 7(b)). Electrical power for these systems was supplied by classical solar panels, of surface area close to 15 square metres, which provided an uninterrupted source of 620 watts—almost 10 times as large as that available to the lander. The communications between the lander, orbiter, and Command Control on Earth were two-fold: the Earth possessed a two-way radio link with both components of the Viking spacecraft for transmission of commands as well as for direct receipt of scientific data; while, in addition, a one-way radio link existed from the lander to the orbiter for data transmission and engineering telemetry. The principal scientific results received through all these channels since the late summer of 1976 and their significance for the furtherance of our knowledge of the Martian evironment will be discussed in Chapter 5.

All the planetary probes referred to so far will eventually meet the same fate: either they will crash-land (or remain) on the surface of the host planet, or they will be captured by the Sun in a heliocentric orbit. Exceptions to this fate are four probes sent out by man to explore the more distant parts of the solar system—Pioneers 10 and 11 launched in 1972 and 1973 to pay a close call on Jupiter, and Voyagers 1 and 2 of 1977 to reconnoitre the planets beyond. Inasmuch as their targets lie outside the domain of the terrestrial planets, we shall have little to say about them in this book. However, one fact cannot be ignored: these spacecraft are destined eventually to leave the solar system, and thus constitute the first man-made contributions to the particulate contents of interstellar space.

A long time will elapse before this will become an accomplished fact. In the meantime, the two Pioneers are en route to Saturn. If nothing untoward happens during their fly-by past this second largest planet of the solar system, they will eventually leave the system along a trajectory directed towards the region of the sky where at night we can see the star Aldebaran. In fact, it will take the Pioneers some 30 million years to cover the distance to Aldebaran (53 light years). However, the star will not wait for the rendezvous but will remove itself to a different part of the sky. The odds are overwhelming that, once the Pioneers leave the solar system, they will encounter no other body in space and will survive in their present form for the lifetime of our Galaxy or more. Their present velocities will not allow them to leave the Galaxy to get lost in intergalactic space; however, it is quite probable that they will outlast the Earth. Isn't this a sufficient reward for having to be sent out on such an inhospitable journey through the dark and cold of interstellar space?

3 The Moon: Our Nearest Celestial Neighbour

'Gentlemen, all of you have doubtless seen the Moon—or at least heard of her.' With these words, according to Jules Verne, President Barbicane opened his famous address to the Baltimore Gun Club on the feasibility of firing a projectile to the Moon. Today, only little more than a century after that fictitious meeting, such a preamble is hardly necessary for this chapter, for television—if not telescopes—has brought the Moon into the homes of hundreds of millions of people. The lunar landscape is now a familiar sight to young and old alike, who, on several past occasions, have watched astronauts on reconnaissance missions ramble along on the surface of this alien world in search of their objectives.

To astronomers, and many others, the Moon has of course been a friend of long standing, and at least a rudimentary knowledge of its motion in the sky goes back very far in the memory of mankind. Since prehistoric times, the waxing and waning of lunar phases† have provided the first astronomical basis for the reckoning of time. Whenever we go back sufficiently far in the history of any primitive civilization, we find its people dependent on the lunar, rather than the solar, calendar. The month became a unit of time long before the year, and the Moon—as the graceful carrier of this knowledge—thus gained entrance as a female deity into the Pantheons of many ancient nations: the Semitic Ishtar, or Egyptian Isis, Tanit of Salammbo, or the nimble-footed Artemis of the Greeks all bear witness to its cult. Although the mythical element has gradually declined, in the more recent past the influence of the Moon on art and poetry has continued unabated. Many of us may recall from childhood the graceful tale of Princess Moonbeam and the Chinese Emperor; while, in more mature years, some of us may have fallen under the spell of the sublime music of Beethoven's *Moonlight Sonata*. A list of other examples of inspiration which the silvery disc of ancient Selene has exerted on art could be prolonged almost indefinitely.

Unlike impressionable artists, astronomers are not always over-appreciative of moonlight or enamoured by its poetical qualities. By this we

†The phase is that fraction of the lunar or planetary disc illuminated by the Sun (full phase = full moon; zero phase = new moon).

do not mean only stellar astronomers, whose professional interests lie far beyond the confines of the lunar orb and who tend to regard moonlight as nothing but an unmitigated nuisance, but also those concerned primarily with the Moon and its environment. They treat moonlight in a manner reminiscent of a medieval torture chamber which, to a layman, may amount to a shocking lack of respect: the light collected by the telescopes is made to pass through prisms or bounce off gratings to be decomposed into a spectrum; it must reduce silver grains on photographic plates or heat the joints of thermocouples; and, worse still, by the intervention of a photosensitive surface it has to be converted into an electrical current to operate recording instruments. This is what an astrophysicist needs to do in order to extract from the light captured by his telescopes the information he desires; he will not be satisfied until he has converted it all into silver grains spotting photographic plates, or into various curves recorded by a pen driven by photocurrents.

In contrast with those who study the light from celestial bodies, astronomers concerned with the motion of the Moon in the sky are, on the whole, a very different lot. With modern techniques both position and time can be measured to a very high degree of accuracy and, moreover, the physical laws of nature relating the Moon's position to time are well known and understood. As a result, the scholars concerned are of the austere type, always know what they are doing, and their professional attention is confined, in general, to the last decimals of their results. They regard the motion of the Moon in the sky as a supreme challenge to their skill and ingenuity. Newton, Euler, Lagrange, Laplace, Gauss and Poincaré—to name only the greatest contributors to its study in the eighteenth and nineteenth centuries—have provided the basis on which precise lunar timetables were made, and thus helped to pave the way to the Moon.

What did they tell us that made it possible for men to visit the Moon? The first task was to establish the *distance* separating us from our satellite. In fact, its value, or at least a close estimate, was determined by Aristarchos or Hipparchos from the relative duration of the successive phases of lunar eclipses more than 20 centuries before our time. More recently, astronomers have determined the distance with much greater accuracy by triangulation and have found it to vary between 356 000 and 407 000 km in the course of each month. The mean distance to the Moon amounts to 384 400 km, which is equal to 60·27 times the Earth's equatorial radius or 0·257% of an astronomical unit. Furthermore, it is less than 1% of the distance to our nearest planetary neighbours, Venus and Mars, even at the time of their closest approach. During his lifetime, many a terrestrial traveller has probably totalled a greater mileage by car than would be involved in a round-trip to the Moon. The astronauts of 1969–1972 used rockets travelling faster than cars and crossed the Earth–Moon distance after free flights lasting approximately 70 hours. Their messages sent by radio from the Moon reached us with a time-delay of only 1·28 seconds, this being the time light takes to reach the Earth from our nearest celestial neighbour.

We may add that the results obtained by celestial triangulation have now been superseded in accuracy by the timing of radar and laser echoes of signals sent out from the Earth. The first radar contacts with the Moon were established from both Europe and America in 1946 and the first laser contact in 1962. Since 1969, after different spacecraft of American, as well as Russian origin (Apollo 11 and 12; Luna 17 and 21) had installed on the Moon cube-corner reflectors (see plate 8) from which laser signals could be returned with greater efficiency, laser contacts with the Moon have been maintained almost continuously. The times of the returning echoes (which can be measured in nanoseconds (10^{-9} s)) have made it possible to determine the instantaneous distances between the signal transmitter on Earth and the reflecting element of the lunar surface with a precision of the order of one metre, about 1000 times higher than that attainable by astronomical triangulation. In other words, measurements of *time* are now leading those of *position* by three orders of magnitude but, due to technical advances in different kinds of measurements, the consequences of so spectacular a breakthrough are still far from being fully explored.

Motion of the Moon: Months and Eclipses

The relative *orbit* of the Moon around the Earth is only approximately an ellipse (of mean eccentricity $e = 0.0549$), because the gravitational attraction of the distant but powerful Sun distorts its shape. So large are the effects of some perturbations arising from this cause (evection) that they were already noted by Hipparchos as early as the second century BC, and the existence of several others (variation, annual inequality) was detected by visual observers such as Tycho Brahe in the pre-telescopic era. The plane of the Moon's orbit is inclined to the ecliptic, that is, the plane in which the Earth revolves around the Sun, by $5°8'43''.4$. Moreover, the period (a *month*) in which the Moon completes a revolution around the Earth and returns to the same position in the sky—the so-called 'sidereal month'—is equal to 27·321 661 mean solar days or 27 days, 7 hours, 43 minutes and 11·5 seconds. Knowing the size, form and duration of the Moon's orbit around the Earth, we can easily establish that the mean velocity of its relative motion averages 3681 km h^{-1}, or about 1 km s^{-1}, corresponding to a mean angular velocity of about 33' per hour, a rate more than sufficient to displace the Moon in the sky by its own apparent diameter in the time of one hour.

We should, however, also bear in mind that in the course of one sidereal month the Sun will move approximately one-twelfth of its entire circle in the sky, a distance which it will circumnavigate in one year or 365·256 358 mean solar days. Therefore, after the lapse of one sidereal month the Moon will not have yet returned to the position in which it would show the Earth the same phase in solar illumination. The time interval P_s between two successive identical phases of the Moon—the so-called 'synodic month'—is thus longer than the sidereal month and is defined by the equation

$$\frac{1}{P_s} = \frac{1}{27 \cdot 321\ 661} - \frac{1}{365 \cdot 256\ 358}, \tag{3.1}$$

which yields $P_s = 29 \cdot 530\ 588$ mean solar days or 29 days, 12 hours, 44 minutes and 2·8 seconds. This represents the time interval between, say, two successive first (or last) quarters or full moons, and was already known by Hipparchos correctly to within one second in the second century BC.

However, if we re-define a month as the time taken by the Moon to return to the same relative place in its orbit around the Earth, this so-called 'anomalistic month' will likewise be longer than the sidereal one (because of the secular advance of the apsidal line of the lunar elliptical orbit), and equal to 27·554 550 mean solar days or 27 days, 13 hours, 18 minutes and 37·4 seconds. Finally, the so-called 'draconic month', the time interval between two successive transits of the Moon through the nodes (i.e. the points of intersection of its orbit with the ecliptic), is the shortest of them all (because the nodes recede) and equal to 27·212 220 mean solar days or 27 days, 5 hours, 5 minutes and 35·8 seconds. This draconic month is of considerable significance in connection with the well known phenomena of solar or lunar *eclipses*.

Who has not seen such eclipses—partial or total—and not been impressed by their effects? As is well known, solar eclipses occur if the Moon at 'new' phase happens to interpose itself partially or totally between the Sun and certain parts of the terrestrial surface. If half a revolution later (i.e. at 'full' phase) the Moon enters the shadow cone cast by the Earth into space in solar illumination, it is seen to be partially or totally 'eclipsed' from any part of the terrestrial surface where the Moon happens to be above the horizon. The occurrence of either partial or total eclipses obviously depends on the relative position of lunar nodes on the ecliptic. If the 'synodic' and 'draconic' months were identical, each new moon and full moon would occur in the same relative position with respect to the nodes. Therefore, we should either have a solar eclipse at each new moon, and a lunar one at each full moon, or no eclipses at all.

We know very well that this is not the case, a fact which in itself is sufficient to prove that the positions of the lunar nodes are not immutable in space. However, it so happens that 223 synodic months are almost equal to 242 draconic months, the difference between the two multiples amounting to only 51 minutes and 41·2 seconds. Thus the positions of the Moon relative to the nodes of its orbit should be almost the same every 6585 days or 18 years and 10–11 days, depending on whether this interval contains four or five leap years. In consequence, should an eclipse of the Sun or the Moon occur at a certain date, it should recur at the same place after 6585 days, a period already known to the Chaldeans two and a half thousand years ago and described by the Greeks under the name of *Saros*. Even closer is the coincidence between 716 synodic and 777 draconic months, which leaves a discrepancy of only 9 minutes and 46·1 seconds. Eclipses recur, therefore, more closely in the same place after 21 144 days, or just under 58 years; time intervals longer still exist for which this case is even more accurate.

The phenomena of eclipses of the Sun or the Moon, especially those which happen to be total, have always attracted man's attention, and reports of them have come down to us from the very dawn of human civilization. Even if we confine our attention only to eclipses of the Moon, we may note that a number of them are known to have influenced historical events: such was the eclipse of 413 BC which, according to Thucydides, caused panic in the Greek Navy off the coast of Sicily and thus affected the outcome of the Peloponnesian War; or the eclipse of 330 BC which enabled the historians to date the battle of Arbil in which Alexander the Great defeated Dareios, the last Persian King of the Archamenid dynasty. The lunar eclipse of 21 November 167 BC occurred on the eve of the battle of Pydna in which the Roman consul Aemilius Paulus destroyed the might of Perseus, the King of Macedonia; this eclipse, as we learn from Titus Livius, was interpreted by the Roman tribune Sulpicius Gallus as a favourable omen. And, more recently, there was the memorable lunar eclipse on 21 February 1504 AD—visible in the West Indies—whose prediction (albeit with the aid of the tables of Regiomontanus) saved the lives of Christopher Columbus and all the crew of his fourth transatlantic voyage from hunger and unfriendly Indians.

Since these heroic days, predictions of eclipses have become a matter of routine and their occurrence has ceased to disturb our equanimity. Moreover, a deeper understanding of the tidal evolution of the Earth–Moon system has disclosed that the phenomena of eclipses are bound to have undergone a profound change in the course of geological history. Cosmically speaking, the time of at least total eclipses of the Sun by the Moon will soon be over for, as the Moon gradually recedes from the Earth at a slow but inexorable rate (caused by tidal friction, on which more will be said in the last section of this chapter), the size of the apparent disc of the Moon in the sky will diminish proportionally. As a result, the Moon will not be able to cover the disc of the Sun completely and the present total occultations will become annular transits.

Physical Properties of the Moon: Rotation, Size and Mass

For the moment let us leave these vistas of the distant future and return to the Moon as we see it today. Even in ancient times the mean apparent diameter of the lunar disc had been known to be close to half a degree (or, for Archimedes or Hipparchos, to a '720th part of the zodiac'). Its more accurate mean value is $1865''.2$, oscillating by $\pm 102''.4$ between the perigee and apogee of the lunar orbit in the course of each month.

Strictly speaking, its radius of curvature of $932''.6$ refers to that sunlit part of the Moon's disc visible at any particular phase. The Moon can be seen as a complete circular disc, free from the phase effect, only when the radius vector of the relative orbit of the Moon around the Earth coincides with that of the Earth's orbit around the Sun (i.e. at the moment of the central eclipses of the Sun or the Moon). In other words, the Moon can be exactly 'new' only at the moment of maximum central eclipse of the Sun (be it total or annular) and exactly 'full' at maximum central eclipse of the Moon (when the only light reaching it is sunlight diffracted through the aureola of the terrestrial atmosphere). Therefore, no moon lit directly by the Sun can be 'full', but is bound to suffer from a small but definite phase effect; the same is true of a 'new' moon as well. Not every full moon is equally full, or new moon equally new. Their phase defects will depend on the angular distance between the Sun to Earth and Earth to Moon radius vectors at any particular time. It may be added that while we are denied a view of a perfectly full moon in direct illumination by the Sun, we can still 'see' it in man-made illumination by radar signals. Pictures of the Moon reconstituted from such signals look quite different from those obtained in natural light, for no object can cast a shadow in radar illumination and contrasts between light and dark spots are caused by the different roughnesses of the ground at radar frequencies. In addition, low-frequency signals also penetrate much deeper into the ground and are backscattered by layers well below the surface visible in ordinary light.

The mean apparent radius of $932''.6$ of the lunar disc at its mean distance of 384 400 km from the Earth corresponds to a mean radius of 1738 km for the lunar globe. Departures of the actual lunar surface from a sphere of this radius do not even locally exceed ± 5 km (see figures 1 and 2) and are geometrically quite complicated. The Moon is therefore essentially a spherical globe, a little more than one-quarter the size of the Earth. Its surface covers an area of just under 37·96 million km^2 and its volume amounts to 21·99 billion (10^9) km^3, or approximately 2·03% that of the Earth.

The entire surface of the Moon is not visible from the Earth because the revolution of the Moon around the Earth (and with it, around the Sun) is not the only motion performed by our satellite. It also *rotates*, about an axis fixed in space, in *exactly* the same period as it revolves around us (i.e. in one sidereal month) and thus almost always shows us the same face. But not precisely so, and the reasons why we do not always see the same face (a fact discernible with the naked eye) are worth a few words of explanation. The first and most important fact is that although the rate of the Moon's axial rotation is uniform to within at least 1 part in 10^6, the angular velocity of its revolution in an elliptic orbit varies (in accordance with

Kepler's second law of planetary motion) with the inverse square of the radius vector, sometimes being ahead and at other times lagging behind the velocity of axial rotation. The difference between the two can cause angular displacements of positions on the lunar surface by as much as 7°54′ as seen from the Moon's centre, a phenomenon detected by Galileo Galilei and known as the lunar 'optical libration in longitude'. Secondly, the lunar axis of rotation is not exactly perpendicular to the orbital plane of our satellite, but deviates from this position by 6°41′. Therefore, sometimes we can see more of one polar cap of the Moon than of the other, and at other times the opposite is true, both in the course of each month. This phenomenon gives rise to an 'optical libration in latitude' amounting to ±6°41′. Finally, when an observer on Earth sees the Moon rise, by looking a little over its upper limb he can see more of those parts of the Moon than would be visible if he were situated at the Earth's centre; when the Moon sets the opposite is true. This 'diurnal libration' (not of the Moon but of the observer) can attain 57″.1 (i.e. the equatorial angular radius of the Earth as seen from the Moon's centre) and is called the Moon's 'mean horizontal parallax'. The diurnal libration superposed on all other librations enables us to see considerably more than one-half of the lunar surface; altogether 59% of it can be seen from the Earth at one time or another and only 41% remains permanently invisible; 18% is alternately visible and invisible.

In October 1959, the cameras aboard the Russian Luna 3 unveiled for the first time the principal features of 18% of the Moon's far side. The remaining 13% was revealed in July 1965 when the Russian Zond 3 (cf table 1) succeeded in recording all but a small fraction of it. Between 1966 and 1968 the entire near and far sides of the Moon were rephotographed with superior resolution by the five American Lunar Orbiters (table 3). Thanks to their work, we are now in possession of almost as complete a coverage of the entire lunar surface as we have for the Earth.

Next to its size, the second physical characteristic of the Moon which is of fundamental importance for its structure is its *mass*. The mass of the Moon, like that of any other celestial body which cannot be placed on our terrestrial scales, can be determined only from the effects of its attraction on another nearby body of known mass. In the case of the Moon, this has first been the Earth and then, in the more recent past, different types of space probes. The mass of the Moon accelerates the free fall of the hard-landers or deflects the trajectories of fly-by spacecraft, and both these phenomena can be measured by radio tracking with a high degree of accuracy. The ratio $m_\oplus/m_{\mathbb{C}}$ of the Earth/Moon masses was found from such work to be equal to 81·302 (correct to within 1 part in 10^5) and since the mass $m_\oplus$ of the Earth is known to be equal to $5\cdot976 \times 10^{27}$ g (cf Chapter 8), that of the Moon comes out to be $m_{\mathbb{C}} = 7\cdot350 \times 10^{25}$ g, or over 73 trillion (10^{12}) tons.

This is the weight of our nearest celestial neighbour, and although it may loom incomprehensibly large for the reader accustomed to terrestrial weights and measures, on a cosmic scale it constitutes only a relatively modest lump. And nor is the mean *density* of the lunar globe unusual: if we divide the lunar mass by its volume of $2 \cdot 199 \times 10^{25}$ cm^3, we find its mean density ρ_m to be equal to $3 \cdot 34$ g cm^{-3}, that is, about the same as that of basaltic rocks in the Earth's crust, and considerably smaller than the mean density of the terrestrial globe ($5 \cdot 53$ g cm^{-3}). The mean gravitational *acceleration g* on the lunar surface is, accordingly, equal to only 162 cm s^{-2}, which is less than one-sixth of its terrestrial value. The mean *velocity of escape* from the lunar gravitational field (i.e. the velocity relative to the Moon which a particle must exceed if it is to escape from the Moon into space) is close to $2 \cdot 38$ km s^{-1}, in contrast with the terrestrial value of $11 \cdot 2$ km s^{-1}. The relatively low gravitational acceleration g prevalent over the lunar surface means that much less work is required to lift weights there, or to throw stones to a distance. Moreover, lunar excursion modules (see colour plate 4) need not acquire a velocity much in excess of 2 km s^{-1} to embark on their homeward journey to the Earth. However, a reduced value of g would also imply that our own weight works less effectively if we wish to hammer anything on the Moon, or drive a shovel into the ground by stepping on it.

Internal Structure of the Moon

For the moment let us abandon the global properties of the Moon and turn our attention to its interior. The primary clues to its structure are already in our hands, namely, its known mass and size combining in a mean density of $3 \cdot 34$ g cm^{-3}. The average density of lunar rocks brought back from the Moon between 1969 and 1972 by the Apollo 11–17 missions ranges between $3 \cdot 1$ and $3 \cdot 5$ g cm^{-3}, virtually the same as the mean density of the Moon as a whole. This coincidence becomes even more convincing when we recall that densities of $3 \cdot 1$–$3 \cdot 5$ g cm^{-3} are those of their material at zero pressure, while the mean density of $3 \cdot 34$ g cm^{-3} refers to matter compressed by self-attraction. These facts preclude the existence of any large differentiation of lunar material in the interior which would be accompanied by an appreciable change of density; thus, to a good approximation, we can treat the lunar interior as a regime which is physically homogeneous.

If so, however, the pressure P inside a self-gravitating homogeneous configuration of lunar mass and dimensions at a distance r from its centre should be given by

$$P = \frac{2}{3}\pi G \rho_m^2 \left(r_{\mathbb{C}}^2 - r^2 \right), \tag{3.2}$$

where ρ_m denotes the mean density of $3 \cdot 34$ g cm^{-3} and $r_{\mathbb{C}}$ the radius of the Moon of $1 \cdot 738 \times 10^8$ cm; $G = 6 \cdot 668 \times 10^{-8}$ cm^3 g^{-1} s^{-2} is the value of the gravitation constant. On the surface ($r = r_{\mathbb{C}}$) of the Moon the pressure should be zero and rises with increasing depth to the central value (at $r = 0$) of

$$P_c = \frac{2}{3}\pi G \rho_m^2 \, r_{\mathbb{C}}^2 = 4.71 \times 10^4 \text{ dyn cm}^{-2} \tag{3.3}$$

or 47·1 kilobars. Such pressures are known to be exceeded inside the terrestrial mantle at depths of only 150 km and can be readily attained in the laboratory. Density changes of silicate rocks under these compressions have been measured and the result discloses that a homogeneous globe of silicate rocks of lunar mass and dimensions should, at zero pressure, possess a mean density of only 3·28 g cm^{-3}, which a central pressure of about 47 kilobars should increase to 3·41 g cm^{-3}.

The pressure inside the Moon should, therefore, be of the order of 10 kilobars throughout most of its mass. But this exceeds the crushing strength of typical lunar rocks, and hence their crystalline structure should be expected to yield to hydrostatic pressure (thus justifying the above expression for the pressure P). Given only a sufficiently long time, albeit one short in comparison with the age of the Moon, the lunar globe should, in accordance with the principles of hydrostatics, settle down to a form of minimum potential energy, which is a sphere. This is why not only gaseous bodies like the Sun or the stars, but also solid celestial bodies like the planets are spherical (or as nearly so as permitted by their axial rotation). The maintenance of a non-spherical shape is a privilege of only the minor celestial bodies—such as small satellites, asteroids or meteorites—whose masses exert insufficient self-attraction to overcome the molecular forces of solid state.

That the crust of the Moon must possess much greater rigidity than that of the Earth (and must have maintained it throughout the astronomical past of the Earth–Moon system) is evidenced by its ability to support mascons in their exposed sub-surface positions—a feat impossible for the Earth where mascons are not present at all. If the age of mascons on the Moon is of the order of 10^9 years (as estimated in a later section in this chapter), the layers supporting them must be at least 1000 times more rigid than those on the Earth down to a considerable depth. That this is indeed so has, more recently, been attested by the combined output of seismometers installed on the Moon by the Apollo missions between 1969 and 1972 (see plate 7). The evidence they provided on *moonquakes*—registered in the thousands since 1969—has given us extensive information on the seismic activity of the lunar globe.

Apart from a general seismic background of a very low level, moonquakes may be divided into three distinct classes:

(i) occasional moonquakes with low epicentres†, triggered by meteoritic or man-made impacts;

(ii) tectonic moonquakes—often recurrent—emanating from hypocentres 600–900 km below the lunar surface; and

†The epicentre is the location of a tectonic quake (moonquake, earthquake) on the lunar or planetary surface.

(iii) deep-seated moonquakes recurrent in the period of one anomalistic month or one-half of it, connected with the principal bodily tides raised on the Moon by the Earth.

In addition to the monthly (or fortnightly) recurrence of moonquakes related to the Earth tides, long-term variations in seismic activity of a given focus—the most notorious one being some 800 km below a point specified by the lunar surface coordinates 28°W 21°S—can also be correlated with fluctuations of the tide-generating field of force. Hence, the tidal strain not only acts as a trigger in activating the moonquakes, but also provides at least a major part of the source of their energy. The requisite

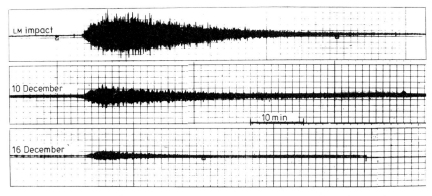

Figure 6. Seismic 'signatures' of lunar events caused by the crash-landing lunar module (LM) (upper panel) and by natural moonquakes (lower panels) on 10 and 16 December 1969, and recorded by the Apollo 12 seismometer. Note the very slow decay of seismic after-effects.

energy is in fact very small in comparison with terrestrial standards. Apart from those produced by the impact of spacecraft, most moonquakes do not exceed magnitude 1–2 on the Richter scale and entail an energy release equivalent to an explosion of about 1 kg of TNT or less per single event. It is only because the Moon is seismically so quiet, dissipating no more than some 10^{15} ergs of seismic energy (equivalent to the explosive energy of about 200 tons of TNT) per annum—in contrast with 5×10^{24} ergs for the Earth—that moonquakes so small and originating so deep can be recorded on the surface at all.

The seismic 'signature' of the observed moonquakes (see figure 6) is very different from that of terrestrial earthquakes. Whereas seismic disturbances on Earth are damped out in a matter of minutes, the aftermath of lunar tremors persists for 60–100 minutes or even longer. Such long echoes can be caused only by an extraordinary amount of scattering of seismic waves in highly fragmented surface layers (about 20 km deep) in which compressional waves propagate with a velocity of about 7 km s^{-1}. Below approximately 60 km, this velocity increases quite abruptly to 8·1 km s^{-1} (no doubt as a result of a corresponding change in composition), and remains the same down to at least 1000 km. However, an analysis of seismic data furnished by moonquakes with epicentres on the Moon's far side has shown that, for seismic events emerging from depths greater than 1000 km (i.e. less than 700 km from the Moon's centre), shear waves cease to leave any discernible trace in seismic messages recorded on the surface. This signifies, in turn, that the lunar material surrounding the centre ceases to behave as an elastic solid and tends to become plastic†. The zone of recurrent moonquakes from which both pressure and shear waves are still received appears to be located just above the level below which shear waves begin to peter out.

The seismic phenomena described above leave no room for doubt that our initial assumption of a homogeneous Moon cannot be true in detail and that the internal density is likely to change with each change in the velocity of propagation of seismic waves. We also possess irrefutable evidence that the distribution of mass inside the Moon is not strictly radially symmetrical. A sufficient proof that the Moon does not conform exactly to a spherically symmetrical model has been provided by the familiar phenomenon of synchronism between rotation and revolution of our satellite—in brief, by the fact that the Moon continues to show us roughly the same face. If its mass were distributed concentrically throughout its interior, its gravitational potential would be equivalent to that of a mass point; but if this were so, the terrestrial attraction would be powerless to exert any effect on its motion and the Moon could rotate at any speed about any arbitrary axis.

But we know that this is not the case: the gravitational effects of the Earth also dominate the motion of the Moon about its centre of mass. In order to give itself up to the mercy of the terrestrial gravitational field, the Moon must have offered the Earth a 'lever' to act upon in the form of different moments of inertia about different axes. The extent of these differences was, in turn, disclosed by the 'physical librations' of our satellite around its principal axes of inertia. These have nothing to do with the 'optical librations' which were described earlier and which are large enough to be detected by the naked eye. The amplitudes of the physical librations of the lunar globe are much smaller and centuries of telescopic observations (augmented more recently by laser links) were necessary to establish their values with the requisite precision. The effort was well worth it, however, for it disclosed that the internal structure of the Moon departs from the requirements of hydrostatic equilibrium (i.e. a spherically concentric distribution of matter) to a small but significant extent. It is clear that the

†Elastic bodies possess the ability to transmit transversal (shear) waves as well as longitudinal (pressure) waves caused by tectonic quakes. In plastic state the body in question ceases to be able to transmit shear waves.

regions where the Moon does so cannot be very far from its surface, for the deeper regions not only become ineffectual for this purpose, but also would tolerate less readily any departures from hydrostatic equilibrium. An outer shell of the Moon 200 km in depth contains 31% of the Moon's mass but accounts only for 46% of its moments of inertia. Below such a depth the lithostatic pressure† exceeds 10 kilobars, a pressure at which rocks can still react as solid in response to disturbances which are short-lasting—such as a passage of seismic waves for instance—but can no longer permanently behave as rigid. In other words, departures from hydrostatic equilibrium, as indicated by the physical librations of our satellite, can maintain themselves indefinitely only in the outer zone of its globe where the lithostatic pressure has diminished to well below 10 kilobars, for beneath this zone the prevalent pressure would eliminate them in a relatively short time. The observed departures are also much too large to be explained by the presence of the individual mascons; their cause must be global rather than local.

This fact entails an important consequence for the past history of the Moon: *it virtually rules out the possibility that the outer zone of the Moon could ever have been molten*. If the Moon was ever covered by a global ocean of fluid magma 200 km or so in depth, it could not have solidified to acquire its present characteristics at *any* distance from the Earth, for these properties do not correspond to the conditions of hydrostatic equilibrium (which would have been preserved by petrification) in *any* gravitational field to which the Moon could have been exposed in the course of its long astronomical past.

In the preceding section of this chapter we arrived at a similar conclusion from the detailed shape of the lunar surface established by the laser altimeters of the Apollo 15–17 missions in 1971–1972 (see figures 1 and 2). This shape also does *not* correspond to any surface of hydrostatic equilibrium, at any distance from the Earth. One could perhaps argue that the shape of the Moon may have been disfigured, after solidification, by intensive meteoritic abrasion in the first few hundred million years of its existence. However, the moments of inertia about the principal axes of the lunar globe constitute volume, rather than surface, characteristics, and could not have been influenced to the observed extent to remain out of harmony with hydrostatic equilibrium.

If, as we believe, the Moon originated by accretion of pre-existing particles in solid state, some lack of radial symmetry of the finished product would, in fact, be more natural than a strict adherence to the principles of hydrostatic equilibrium. In the long run, the latter asserted themselves throughout most of the deep interior of our satellite. Its present shape and the moments of inertia of its outer zone—where self-attraction was not sufficiently strong to enforce hydrostatic equilibrium—may still reflect the vagaries of the formation of the Moon in its last stage of accretion.

†Lithostatic pressure is the equivalent in solids of hydrostatic pressure in liquids.

The fact that most igneous rocks brought back from the lunar surface are thermally differentiated is not really at variance with these conclusions. Dynamical arguments advanced in their favour do not exclude *local* melting of the lunar surface—on the size of the maria or large craters—as basins where differentiation could be accomplished, if indeed such a differentiation of material occurred on the finished Moon and not in its formative stage. But they do rule out the existence, at any time in the past, of a global ocean of molten magma covering the whole Moon; such a hypothesis conflicts with well established dynamical facts from which there is no escape.

The *temperature* of the lunar surface at the present time is controlled wholly by insolation. The heat flux† at depths beyond the range of the diurnal heat wave (but not more than 1–2 metres) was measured only twice by the Apollo missions: firstly, in the proximity of the Hadley rille by Apollo 15 in July 1971; and secondly, near the Taurus–Littrow landing place of Apollo 17 in December 1972. The results obtained may, however, be of only local significance and do not add up to give a consistent picture. On the other hand, the disappearance of shear waves from the seismic signatures of moonquakes from depths exceeding 1000 km leads us to conclude that, at these depths, the temperature of rocks may have exceeded 1200–1400°C to make them behave as plastic under the prevalent pressure. The Moon's centre is probably no hotter than 1500°C.

Another constraint on the actual temperature of the Moon's interior has been provided by the observed magnetic interaction of the lunar globe with the 'solar wind' (a hot plasma evaporating from the solar corona and carrying its own magnetic field through space) or rather the lack of it. A systematic study of such an interaction commenced in 1967 with the launch of Explorer 35, a lunar satellite especially instrumented for this purpose (see table 3). Magnetometers aboard this space probe have demonstrated that, unlike the Earth, the Moon does not produce any shock waves in the interplanetary medium through which it orbits with supersonic speed. Instead, it casts a geometrical shadow in the anti-solar direction and behaves as an insulator rather than as a semiconducting body. An upper limit for the electrical conductivity of the Moon which would be consistent with these measurements was found to be of the order of 10^{-5} mho m^{-1}. On the other hand, the electrical conductivity of lunar rocks is a known function of temperature. If we are to reconcile the physical properties of lunar rocks with the magnetometric evidence supplied by Explorer 35 and subsequent space probes, we are led to a conclusion that the temperature inside the Moon at a depth of 700–800 km does not exceed 1000°C, a temperature which is sufficiently low for the respective layers to possess a degree of rigidity consistent with the seismic data.

The work of Explorer 35 and other space probes also provided

---

†The amount of heat received by the surface per unit area and per unit time.

evidence on the general *magnetic field* of our satellite. That any such field which the Moon may possess must be very weak had already been revealed by the experiments performed by the Luna 2 space probe in September 1959, a result which was fully confirmed by all the subsequent work. Consequently, we now know that the general dipole field of the Moon cannot exceed a few gammas (i.e. 10^{-5} gauss) in strength, and that the total magnetic moment of the lunar globe is probably less than one-millionth of that of the Earth.

With the arrival in the lunar environment of the surface and sub-satellite magnetometers of the successive Apollo missions, many new facts on lunar magnetism came to light. The sub-satellites of the Apollo 15–17 missions orbited at altitudes substantially lower than that of Explorer 35 and detected local magnetic fluctuations between 20 and 30 gammas. These fluctuations turned out to be correlated with specific formations (craters) on the lunar surface overflown by the magnetometer. The carriers of this field are brecciated rocks† produced by impacts and there is a strong presumption that they were also magnetized by impacts. At any rate, the origin of this field which fluctuates with topography can only be skin deep, and has nothing to do with the deep interior of the Moon.

The same is not true of the remanent magnetism of lunar crystalline rocks brought back by successive Apollo and Luna missions. The magnetism of these rocks is stable and suggests a prevalence of much stronger fields (100–1000 gammas) at the time of their solidification. Such fields would be 50 to 500 times less intense than the present magnetic field of the Earth, but 20 to 200 times stronger than the fields carried by the present-day solar wind. Whether or not the mechanism which generated such fields in the past was internal (a dynamo in a liquid conducting core, which has since become inoperative because of solidification), or external (induced by the Earth at closer proximity, or anomalous solar wind) is as yet unclear. Both these alternatives are unlikely for many reasons, but nothing more plausible can be advanced in their place. The origin of the remanent magnetism of lunar crystalline rocks continues to present us with an unsolved problem.

The Surface of the Moon and its Formations

In the preceding sections we became acquainted with the fundamental physical properties of the Moon and of its interior. Our next objective should be to extend this acquaintance to the visible surface of the Moon and to the formations appearing on it. What is so arresting about its face, and what can we learn from it about the past and present state of our satellite?

†Breccias are broken rocks subsequently reheated to melting point and compressed into conglomerates of the originally heterogeneous material.

The Moon: Our Nearest Celestial Neighbour

Plate 10. A view of the Moon photographed from the Earth by the Manchester Lunar Programme at the Observatoire du Pic-du-Midi.

Even to the naked eye the Moon is seen as a beautiful object. Its face is diversified with markings which have been associated with the various popular myths ('the man in the Moon') of many civilizations. If we look at this face through a telescope (or, more comfortably, at the photograph of it in plate 10), we see that its surface consists mainly of two different types of ground. One is rough, articulate, relatively bright (reflecting in places as much as 18% of incident sunlight) and essentially mountainous, replete with hills and other types of vertical formations which we shall describe and identify presently. The other type is darker (reflecting on average only 6–7% of incident sunlight) and much smoother, its plains being occasionally dotted with hills. The first type of ground is commonly named the *continents*; the term 'highlands', which is sometimes used, is a misnomer perpetuated in literature of lighter vintage, for not all continental ground is in an elevated position. The other type, the flatlands, is referred to as the *maria* or 'seas'—an even greater misnomer (going back to Galileo Galilei) since no drop of water ever wetted their surface. However, the term took root in popular parlance, and for this reason we shall continue to use it as well.

The continents and maria cover the surface of the Moon in very unequal proportions. More predominant are the continents, which cover some two-thirds of the Moon's near side and more than nine-tenths of the far side. The maria are, in contrast, located almost exclusively on the near side of the Moon and distributed broadly along its equatorial belt—a fact no doubt connected with the mode of their formation and which we shall discuss later.

A closer telescopic inspection of both the main types of lunar ground reveals that they are covered with an almost limitless number of mountainous formations of all sizes, no two of which are exactly alike. The most characteristic type among them—and by far the most numerous in any part of the Moon—are ring-like walled enclosures commonly called *craters*. This Greek word is used here in its original sense meaning 'cup' or 'bowl' and without prejudice for the views on its origin, for too specialized an interpretation could easily render the word as much of a misnomer as the Martian 'canals' (see pages 125–9) or the lunar 'seas'.

The number of craters on the Moon is indeed immense: those with diameters in excess of 1 km are estimated to more than 300 000 on its visible hemisphere and to at least one million on its far side; those smaller still number too many for any realistic estimate. There are 16 formations of the type enclosed within unbroken walls with diameters in excess of 200 km, although only 5 are in the visible hemisphere. The crater Clavius near the Moon's south pole is one of them, and a photograph of it is reproduced in plate 11. Over 70 craters (32 of which are visible from the Earth) possess diameters between 100 and 200 km. These are usually characterized by fairly smooth floors, sometimes checkered by smaller craters (such as, for instance, Clavius) which are again absent in others. The elevation of their walls (1–3 km) represents so small a fraction of their dimensions that even their rims are mostly below the horizon for an observer situated at their centre. Typical examples of craters on the Moon's near side with diameters between 90 and 100 km are Copernicus (plate 12), Theophilus (plate 13) and Tycho (plate 14). These are characterized by hummocky walls, rough floors, and the frequent presence of groups of hills constituting the 'central mountains' of such formations. In contrast, craters substantially smaller (10–20 km in diameter) seldom, if ever, possess any central peaks and their upturned rims are barely distinguished by their inclination from the surrounding landscape. Finally, crater formations smaller than 1 km in size (the properties of which could only be explored by mooncraft) lack any walls at all, and constitute merely funnel-like depressions with small, flat floors.

Apart from some distinguishing features which depend mainly on size, all lunar craters also possess certain characteristics in common which

Plate 11. A photograph of the lunar crater Clavius near the Moon's south pole, taken with the 200 inch Hale reflector of the Mount Wilson and Palomar Observatories.

are indicative of their origin. In order to enumerate them, let us stress first that the distribution in size of lunar craters ranges continuously from the largest formations to the smallest pits; no size is missing, and all appear to be distributed on the Moon at random. Secondly, the heights of their ramparts are very small in comparison with their dimensions, again largely regardless of size. However—and this is very significant—*their floors are generally depressed below the level of the surrounding landscape* by amounts which may run into several hundred metres for large craters. In fact the larger the crater, the greater the depression ('Ebert's rule') so that, in general, the craters resemble not mountains, but pockmarks dotting the lunar landscape in prodigious numbers. That this is so represents a very eloquent testimony to the *origin* of these formations, for if lunar craters are depressions rather than elevated structures, then this suggests that the force which gave rise to them was directed *downwards* rather than up, and caused *removal* of material instead of accumulation. This all suggests that the *origin of craters*

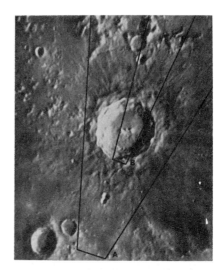

Plate 12. A view of the lunar crater Copernicus (upper right) as photographed from the Earth, in contrast with views recorded by Lunar Orbiter 2 on 23 November 1966 from a close proximity to the target. The photograph in the upper left shows an oblique view of this crater as recorded by the Orbiter's wide-angle camera, while the photograph below shows a high-resolution view of the interior taken from an altitude of 45 kilometres. The fields of view of both the Orbiter photographs are marked on the terrestrial photograph by the cones A and B, respectively. *Photographs by courtesy of NASA and Manchester Lunar Programme.*

Plate 13. Sunset over the group of lunar craters Theophilus (lower left), Cyrillus and Catharina on the eastern shores of Mare Nubium, as photographed with the 24 inch refractor of the Observatoire du Pic-du-Midi (Manchester Lunar Programme). The walls of Theophilus and Cyrillus overlap in a way which demonstrates that Theophilus is the younger of the two.

and crater-like formations on the lunar surface was due to external action and *not* to causes emanating from the interior.

For a long time this was far from certain, and arguments between the internal and external origin of such formations go back to the early days of telescopic astronomy. The first person who seems to have considered both aspects of the problem was Robert Hooke (1635–1703). Since his time the weight of scientific opinion has oscillated between the two arguments until almost the present day. It is true that, in the more recent past, astronomers studying the particulate contents of interplanetary space have pointed out that while internal causes (favoured by many geologists) *may* have co-operated in shaping the face of the Moon, external impacts *must* have been operative in producing the craters, for the Moon does not orbit in empty

Plate 14. A photograph of the crater Tycho taken with the 43 inch reflector of the Observatoire du Pic-du-Midi (Manchester Lunar Programme).

space, and nor is it protected from the impacts of bodies whose orbits happen to intersect that of the Moon.

But the real decision did not come until the first astronauts and their cargo returned from the Moon in 1969, when they informed us not only of the composition of the lunar crust, but also of the type of minerals constituting it. On the Earth, all rocks occurring in the crust belong to one of the following three groups: (a) solidified sedimentary deposits; (b) metamorphic rocks (one-time sediments whose structure was subsequently changed by pressure or temperature); and (c) igneous rocks which crystallized from volcanic magmas (lavas). This last group can, in turn, be subdivided into fast-cooling rocks which exhibit a fine structure of small crystals, and into slow-cooling ones in which larger crystalline structures had a chance to develop. Basalts are typical examples of the former and granites of the latter. On the Moon, rocks of types (a) and (b) are conspicuous by their absence, a fact which attests to a complete absence of water on the Moon not only at the present time, but also throughout its entire past. All the 382 kg of rocks brought back by the successive Apollo missions consist entirely of group (c), that is, igneous rocks, the structure of which discloses that they solidified from molten magma at temperatures between 1100 and 1200 °C under highly reducing conditions. From their atomic compositions, the most abundant element is oxygen (making up 60% of the Moon's crust by weight), followed by silicon (16–17%), aluminium (6–10%), calcium (4–6%), magnesium (3–6%), iron (2–5%), and titanium (1–2%) in progressively diminishing amounts. All other elements are present in amounts very much smaller than 1% by weight. If we compare these abundances with those encountered in the crust of the Earth, the first three elements (O, Si, Al) are present on the Moon in comparable amounts. However, the iron and titanium contents are distinctly enhanced on the Moon, while the alkali metals are less abundant, and so is—by orders of magnitude—carbon or nitrogen.

Of the compounds formed by these elements, quartz (SiO_2) constitutes between 40 and 50% of the Moon's crust by weight, as compared with its 48·5% abundance in the crust of the Earth, while ferrous oxide (FeO) or calcium oxide (CaO) each constitute 10–20%. A significant fact is that all oxidized compounds appear to be present on the Moon only in their *lowest* states of oxidation. The oxide of the lightest metal of all—hydrogen—in the form of H_2O is totally absent on the Moon; no trace of water in any form has been found anywhere on our satellite so far. The only form of hydrogen which must be present on the Moon is that brought in by the solar wind (cf p 78), and any water produced by its oxidation will be quickly dissociated by sunlight.

From the mineralogical point of view, the backbone of the dark crystalline material which fills the basins of lunar maria can be described as 'gabbroid basalts', material akin to lavas known on the Earth but enriched with iron and titanium. In contrast, the continental areas of higher reflectivity appear to consist of feldspathic rocks—including a nearly pure feldspar called anorthosite—which are akin to terrestrial granites. Anorthosites lack the iron and magnesium of basaltic rocks, having replaced them with aluminium (this is what makes them lighter in weight as well as in colour). The very existence of anorthosites on the Moon implies chemical differentiation of the crust, in the course of which the heavier elements, like iron, were separated from the lighter ingredients. Moreover, anorthosites are mostly coarse-grained minerals, which means that they must have cooled off slowly from the melt and scarcely on the lunar surface.

But our main concern with the origin of the lunar surface formations is not so much the chemical (atomic or molecular) composition of the constituent rocks, but their physical texture. Of signal importance in this connection is the fact that the bulk of the material (85–90% by weight) brought back from the lunar continents appears to be formed by *breccias* in which fragments of diverse origin were welded together by events subsequent to their first solidification. The structure of such breccias, with its evidence of shock metamorphism, leaves no doubt as to their origin: *they were produced by impacts of celestial bodies of different size on the lunar surface* which impressed upon it (and, in particular, the continental areas) its characteristic macrostructure. In their present form they certainly never passed through the orifice of any volcano known to us on Earth, and nor have such volcanoes (at least of the Vesuvius type) been found anywhere on the lunar surface on any scale, large or small.

These facts leave no room for doubt as to the nature of the processes responsible for the stony sculpture of the lunar continents or the marial plains: these were caused by external impacts of bodies of all sizes populating interplanetary space (see Chapter 6) which happened to be on a collision course with the Moon. Such a possibility was considered by Robert Hooke in his *Micrographia* (1667), although he rejected it at that time because 'it would be difficult to imagine whence those bodies should come.' We now know better; and the prevalence of brecciated rocks in the lunar surface material established in 1969 has clinched the argument for good.

Cratering of the Lunar Surface

With the certainty of impacts now safely established, let us consider their probable course and consequences. In order to visualize them, consider a moderately large meteorite, of the size and mass of a rock weighing (say) one million tons (10^{12} g), impinging on a solid surface with a velocity v of (say) 30 km s^{-1}, that is, a velocity equal to that of the Earth in its relative orbit around the Sun. The kinetic energy

$$E = \tfrac{1}{2} mv^2 \tag{3.4}$$

of such a missile would be of the order of 10^{25} ergs and would enable it to pierce the lunar crust quite far down before eventually coming to rest. In the course of its deceleration, which lasts only seconds, the initial kinetic energy of the missile must of course be conserved and the entire amount reappear in other forms, mainly as thermal energy, mechanical energy of shock and fracture, and seismic energy of elastic waves.

If, for the sake of argument, the entire initial kinetic energy of the meteorite were to be converted into heat, its temperature T should be given by the equation

$$T = v^2/2C_V, \tag{3.5}$$

where C_V denotes the specific heat of the meteoritic material. As C_V is a quantity of the order of 10^7 erg g^{-1} deg^{-1} for stony meteorites, the temperature resulting from equation (3.5) for $v \sim 10$ km s^{-1} should be of the order of several million degrees, an amount sufficient to vaporize completely the whole impinging mass and convert it into an extremely hot bubble of gas at a depth of several diameters of the original body beneath the lunar surface. In reality, the actual temperatures would of course be considerably lower, because a large part of the original kinetic energy would also be converted into other forms. Nevertheless, it is virtually certain that an impact of such a body would create temperatures of the order of 10^5 degrees for a short time (suffice it to recall that a body moving at a speed of a mere 3 km s^{-1} possesses a kinetic energy equivalent to that released by the explosion of an equal weight of TNT).

Needless to say, a bubble of so hot a gas could not be contained by the weight of the overlying debris for longer than milliseconds. It would immediately expand with great violence, and the effects of such an expansion would severely affect regions that are very large in comparison with the size of the cause of the original disturbance. If the density of the latter had been close to 3 g cm^{-3} (appropriate for stony meteorites), the dimensions of one weighing a million tons would have been a little more than 40 m across. The main effect of its subterranean explosion would, therefore, be essentially that of a point charge and the direction of the intruder could not have much influence on the symmetry of the scar produced on the surface. The probable result of this event is schematically shown in figure 7, which represents the expected cross section of the resulting impact crater. The solid material left by the intruder itself should be negligible, for most of it should have evaporated to escape back into space, or become dispersed over a large part of the adjacent lunar surface.

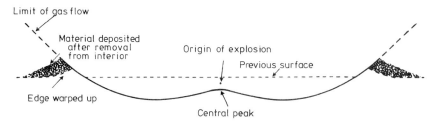

Figure 7. A schematic profile of an impact crater.

Extensive terrestrial experiments ranging from laboratory models to nuclear explosions in the megaton TNT range disclose that the diameter D (in km) of an impact crater formed by a shallow sub-surface explosion should be related to the kinetic energy E (in ergs) of the impacting body by an empirical relation of the form

$$\log D = 0\cdot 29 \log E - 6\cdot 9, \tag{3.6}$$

a relation which can be extrapolated for larger impact formations encountered on the Moon. If so, craters in the 100 km range like Copernicus (plate 12), Theophilus (plate 13) or Tycho (plate 14) could have been formed by impacts of bodies possessing kinetic energies of the order of 10^{30} ergs (corresponding to 10 million megatons of TNT), while a formation of the size of Clavius (plate 11), over 200 km across, would have called for an energy 10 times larger. If, moreover, the velocity of impacts was typically of the order of 10 km s^{-1}, the masses and dimensions of (stony) meteorites possessing such energies should have been of the order of 10^{12} tons and 20–30 km for craters like Copernicus, and 10^{13} tons and 50 km for craters double their size. Such bodies are reminiscent of small asteroids of the size of Adonis, Hermes or Eros, to mention only a few of those which have paid close calls on us in recent decades (see Chapter 6).

The impact of such a body would not only create a scar of the requisite size on the lunar surface, but also would eject a tremendous amount of material at sub-orbital velocities to impact elsewhere on the Moon (predominantly in the neighbourhood of the impact). Since the velocity of escape from the gravitational field of the Moon is only 2·4 km s^{-1} (see p 46), the velocities of these 'secondary impacts' would be typically 10 times smaller than those of the 'primary' one which gave rise to the crater. As a result, the energies per unit mass of the secondary ejections would have been 100 times smaller than that possessed by the cosmic intruder, a fact which should make the dimensions of the 'secondary' craters originating in this way some 100 times smaller than that of the 'primary' crater whose formation gave rise to them. Such a prediction is amply borne out by actual evidence: primary craters like Copernicus are indeed surrounded by hundreds of secondaries of the expected size (apart from other ejecta (bright rays), on which more will be said in the penultimate section of this chapter).

What more can we say about the source of cosmic particles responsible for primary impacts? Unlike Robert Hooke in the seventeenth century, we no longer have to wonder where these may come from. But what conclusions can be drawn about their motion prior to impact from the evidence preserved in the stony relief of the lunar face? Although explosions caused by primary impacts are essentially akin to those of a point charge, the form of the craters so produced can also tell us something about the angle of incidence of the intruder, at least if the impact occurred at low angles (less than 10–15°, say) to the lunar surface. The scars left by high-angle impacts are essentially circular or polygonal, but at low angles of impact they become elongated along the plane of the impact (see plate 15). It is this which can disclose the circumstances.

We have already mentioned that craters of all sizes appear to be distributed over the lunar surface at random and this applies equally to craters of any size or shape, including those produced by low-angle or grazing impacts. But if so, *the impacting bodies could scarcely have been in heliocentric orbits*, for had the inclinations of such orbits to the ecliptic been small, most of them would have missed the Moon altogether. Among those which did not, high-angle impacts should have occurred predominantly in the equatorial zones of the Moon, while low-angle or grazing impacts should be concentrated in the polar regions.

The face of the Moon discloses conclusively that this is *not* the

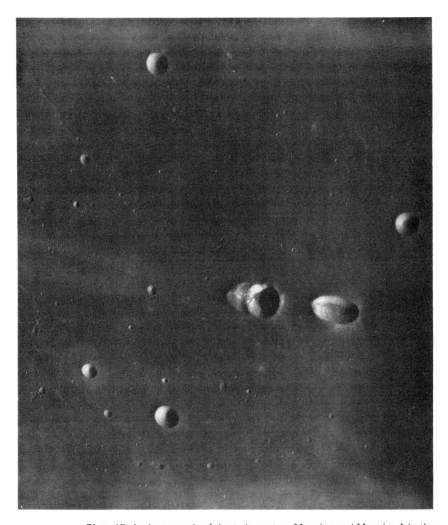

Plate 15. A photograph of the twin craters Messier and Messier A in the western part of the lunar Mare Tranquillitatis taken in the proximity of the lunar surface by the Apollo 11 mission. The forms of these craters are distinctly elongated, suggesting an origin by oblique impact. *Photograph by courtesy of NASA.*

case and that impacts at all angles appear to have occurred at random in all parts. This means that the particles destined to end their independent cosmic careers by collision with the Moon were not revolving around the Sun prior to impact, but were actually within the influence of the Earth–Moon gravitational dipole. In other words, particles responsible for the bulk of the cratering of the lunar surface were probably cosmic left-overs of the process which resulted in the formation of the Earth–Moon system, and were carried with it until they were eventually swept up by these two mass centres. If true, this would imply that the Earth and the Moon were already gravitational partners (though not necessarily at their present distance) during the epoch of intensive cratering of the lunar surface which is thought to have occurred during the course of the first few hundred million years of its existence.

Is there, perchance, any escape from such a conclusion? One way would be to assume that most of the impact craters were inflicted on the lunar continents before the Moon was captured by the Earth. The Moon, travelling alone in space, may have been tumbling about its centre of mass so that heliocentric impacts from all directions could have been equally likely. But this is very improbable. It is true that we have no idea whether or not the Moon was spin-stabilized before its hypothetical capture; its present state of rotation is certainly controlled by the Earth. However, it is most unlikely (cf p 96) that any pre-existing surface sculpture would have survived this brutal experience, and it is hard to avoid the conclusion that all we observe today came into being when the Moon was already a satellite of the Earth.

We shall return to the problems arising in this connection in the last section of this chapter; but here we wish to say a few words on the origin of the lunar *maria*. That such formations—particularly the so-called 'circular maria' whose shores were not modified by events subsequent to their formation—are nothing but gigantic *impact craters* is strongly suggested by their morphological characteristics as shown in plates 16 (Mare Imbrium) and 17 (Mare Orientale). Petrographic evidence in the form of brecciated rocks procured by the Apollo missions has left no room for doubt that this is indeed the case. The diameters of the largest of these maria are of the order of 1000 km and if equation (3.6) could be extrapolated to formations of this calibre, impacts by bodies of 10^{16} tons in mass and 100 km in size would be necessary to create them. These dimensions are characteristic of moderate-size asteroids existing in great numbers in the asteroidal belt of the solar system (cf Chaper 6). Whether any of these could have strayed so far from their principal reservoir still remains an open question.

Incidentally, the ramparts of these and other maria represent the only 'mountain chains' encountered on the surface of the Moon. In the case of Mare Orientale (plate 17) the triple ring of mountains surrounding its central 'eye' is still well preserved on account of its relatively young age (see p 88). Those bordering Mare Imbrium—the Apennines and Alps on its eastern shores, or the Carpathians to the south—together form an incomplete ring, parts of which were destroyed by subsequent events. These ramparts sometimes rise to impressive heights: Mount Hadley in the lunar Apennines attains an altitude close to 5000 m above the eastern shores of Mare Imbrium, an altitude exceeding that of Mont Blanc in the Alps. But—and this should be emphasized—*the origin of such mountains on the Moon has nothing to do with any folding of the lunar crust* or anything to do with 'plate tectonics' so familiar to us on Earth. The lunar mountains were apparently raised to their present elevations by instantaneous catastrophic

Plate 16. A photograph of the lunar Mare Imbrium taken at the time of the last quarter with the 100 inch reflector of the Mount Wilson and Palomar Observatories. The crater Copernicus can be seen near the upper right-hand corner.

events and, in the absence of any significant erosive processes operative on the Moon which could denude them, they retained their altitudes for periods of time which would have been sufficient for complete obliteration had they been on the Earth.

Another important aspect of the difference between the lunar maria and smaller craters concerns their distribution over the lunar surface. While craters smaller than (say) 200 km in diameter are distributed essentially at random over the entire surface of the Moon, the mare plains are decidedly not, being located predominantly on the near side of the Moon at low or moderate selenographic latitudes. If all these plains were formed by impacts, then this means that the impacting bodies crash-landed predomi-

Plate 17. A photograph of the lunar Mare Orientale taken by Lunar Orbiter 4 in May 1967. *Photograph by courtesy of NASA.*

nantly on the hemisphere of the Moon facing the Earth and along its equator. These facts (as well as other circumstances which will be explained in the last section of this chapter) make it probable that *the impinging bodies were once satellites of the Moon* and remained as such until the inexorable laws of celestial mechanics compelled them to crash-land on the lunar surface to produce the greatest scars on its face conspicuous even to the naked eye.

The Lunar Environment

Our acquaintance with the surface of the Moon should not end with a guided tour of the principal types of formations of the lunar landscape as seen by the naked eye or through a telescope. Of equal interest to us should be its small-scale structure and chemical composition. What are the physical and chemical properties of the lunar surface on which the astronauts of 1969–1972 had to perform their appointed tasks? While we know this now, of course, as a result of their experiences, it is of interest to point out that certain essential features of the microstructure of this surface—hopelessly beyond the limits of telescopic resolution from the Earth—could have been inferred, in advance of the space age, from certain observations which can be made with the naked eye—mainly from the variation of light which we receive from the Moon at different phases.

We have already pointed out that the Moon is a pretty poor reflector of incident sunlight; only about 7% of it is scattered by the Moon as a whole. If the surface of the Moon were as light as our terrestrial cloud cover, for instance, moonlit nights would be very much brighter than they actually are, and even our senior citizens could easily read their newspapers outside on such nights! The main reason why this is not so is the fact that the lunar surface is relatively dark and no brighter, on average, than volcanic lava fields on the Earth.

What interests us at the moment is not so much the poor quality of the Moon as a reflector, but rather the variations in the amount of moonlight with the phase. Did you ever stop to note the illumination of a moonlit landscape at different phases of the Moon? If you had measured it, you would have found that, although the full moon occupies only twice as large an area in the sky as it does at its first or last quarter, the night of the full moon is, not twice, but 15–17 times as bright as that of the first quarter was a week before, or the last quarter will be a week later. In particular, within a few hours of the full moon (and remember that not all full moons are equally full; see p 44) a surge in intensity of lunar illumination is truly conspicuous, a fact which could not be true if oblique sunrays were illuminating a smooth surface, no matter what its substance might be. It can be understood only if the *small-scale structure of the lunar surface is highly vesicular (honeycombed)* and capable of very strong backscattering: that is, it begins to cast a shadow on itself when the Sun is still very close to the zenith. This appears to be true not only of the full moon as a whole, but also of the amount of light received from any element of the lunar surface, be it continent or mare. Thus the backscattering ability extends over every type of lunar ground and has nothing to do with its chemical composition.

This observation cannot by itself specify the scale of the roughness of the lunar surface—whether, for instance, it is on a metre, centimetre or millimetre scale. However, other observations can help us to do so. Firstly, the measurements which indicate that the colour of scattered moonlight is essentially the same as that of the illuminating sunlight, that is, 'white' (or, rather, grey); and secondly, the fact that moonlight is partly polarized and its degree of polarization varies with the phase. If the process of scattering is to be independent of colour, the dimensions of the particles responsible for scattering must be large in comparison with wavelength, but again not too large to produce the observed kind of polarization. Dust grains of an average size of 0·01–0·1 mm would meet these requirements. The actual photographs taken by the astronauts on the Moon show the extent to which this reasoning was borne out by reality (plates 18 and 19).

But before we descend to the surface of the Moon with the lunar soft-landers and their crews, let us consider the information which reached us on waves of light before their landings. Earlier in this section we stressed the fact that only about 7% of incident sunlight is *scattered* by the lunar surface to produce the 'silvery moonlight' of our songs and romances. What happened to the rest? It must have been *absorbed* by the lunar surface and converted into heat. But each surface so heated must also *emit* radiation of its own. The intensity of this emission, as well as its spectrum, is in turn governed by the temperature of the emitting surface. For the Sun, whose absolute temperature is a little more than 5700 degrees, most of the radiation is emitted as visible light with a maximum in the yellow, and this is also the colour of reflected moonlight. However, the proper light of the Moon turns out to be of a very different kind.

The sole and sufficient reason for this is the fact that the temperature of the lunar surface is much lower than that of the Sun. Apart from a minute trickle of heat of internal origin, the Moon—like the Earth—receives all its heat by radiation from the Sun. But as their average distance from the Sun amounts to 214 solar radii, each unit area of the lunar surface receives only one $(214)^2$-th part of the heat flux passing through the same area of the surface of the Sun. As this flux should be proportional to the fourth power of the absolute temperature (in accordance with Planck's law), the mean temperature of the Moon should be $(214)^{2/4}$, or about 15 times lower than that of the Sun, which is approximately 380 degrees kelvin (i.e. +107 °C). This should, of course, only be true if the Sun stands in the zenith; should its light fall obliquely on the surface, the temperature maintained by it would be correspondingly lower. But whatever it may be, most of the thermal radiation should have been emitted at wavelengths around 10 μm or 0·01 mm, that is, in the infrared region of the spectrum. Light of these wavelengths is invisible to the human eye and incapable of impressing a photographic plate. It will, in addition, experience considerable difficulty in penetrating our atmosphere through the interlocking absorption bands of water vapour and carbon dioxide. Nevertheless, that part of it which does get through can be detected and measured quite accurately by photoelectric devices, leaving no doubt that we are, in fact, living under an 'infrared' rather than a 'silvery' Moon, regardless of all the poetical fiction of the past.

We may add that the thermal devices taken to the Moon by the lunar soft-landers of 1966–1972 verified the results of the earlier measurements of thermal flux at a distance, although the manned missions never landed on the Moon when the Sun was near the zenith. With the advancing

The Moon: Our Nearest Celestial Neighbour 69

Plate 18. Man on the Moon. This photograph was taken at the Tranquillity Base on 20 July 1969 by Neil Armstrong and shows Edwin Aldrin standing in the desolation of the lunar landscape. Reflected in the visor of his helmet are photographer Armstrong, the American flag, and (on the right) a part of the excursion module *Eagle*. *Photograph by courtesy of NASA.*

Plate 19. Man's footprint on the Moon. The rate of erosion on the lunar surface is so small that (if undisturbed by human activity) this footprint could survive essentially as we see it on the photograph for a time-span of the order of 100 million years. *Photograph by courtesy of NASA.*

time of the day, which on the Moon lasts 29·53 days of our own time, the temperature begins to decline and continues to do so after sunset by cooling throughout the night, until, just before sunrise, it falls to approximately 80 K or -190 °C. Therefore, the diurnal temperature variation in the lunar tropics can attain 300 °C, ranging roughly from the temperature of boiling water (at normal pressure) to that of liquid air. At higher latitudes, where the Sun never reaches the zenith, noon-time temperatures will generally be lower than 100 °C; and near the poles, where the Sun never rises much above the horizon nor sets completely for long, the diurnal temperature variations become correspondingly smaller. The *mean* monthly temperature on the Moon is not too different from that on the Earth as both these bodies receive (per unit area) equal amounts of heat from the Sun. It is the low heat capacity of the Moon's surface material which makes the lunar extremes so large.

In another respect the lunar climatic changes are not as drastic as some of the foregoing figures may seem to indicate. Although the extremes are large, the period of the thermal cycle, which lasts one synodic month or 709 hours of our time, is so long that the rate of change of lunar temperatures seldom exceeds $\pm 1° h^{-1}$. It is only the persistence of such gradients over so many hours of rise and fall that makes the extremes of temperature so wide.

There exists, however, a time when the lunar temperature changes can become very much more rapid. This happens during the relatively brief intervals of eclipses when the Moon passes through the shadow of the Earth. Such events are not too rare: they occur about once a year (or 12 lunar days) and last only a few hours. During this time the eclipsed parts of the lunar surface experience almost as large a variation in temperature as they do between day and night. In particular, the emergence of the Moon from the shadow cone brings about a temperature rise of almost $200° h^{-1}$, a rate observed first from the Earth and then verified on the Moon by Surveyors 3 and 5 in 1967 (see p 14).

If these extremes of temperature may, perchance, give some less courageous individuals second thoughts about spending a month's holiday on the Moon, let us reassure them quickly that such large climatic variations are limited solely to the exposed surface and become very much smaller immediately beneath it. The discovery of this fact actually preceded the space age and represents another gift which the science of the heavens owes to its junior branch of radio astronomy. We have already mentioned that, at surface temperatures on the Moon ranging between 100 and 400 K, most of the thermal emission governed by Planck's law lies in the far-infrared region with a maximum intensity around the wavelength (λ) of 10 μm, falling off steeply towards shorter wavelengths and somewhat more slowly towards longer wavelengths. Most of this 'moonlight' between the wavelengths of 10 and 1000 μm (i.e. 0·01–1 mm) will be irretrievably absorbed by our own atmosphere. But for wavelengths beyond 1 mm (corresponding to frequencies below 3×10^5 MHz) the transparency of the atmosphere improves rather suddenly and enables us to observe the Universe in this strange light even from ground-based facilities.

When radio-telescopes capable of recording such radiation were directed at the Moon, they registered phenomena which, although surprising at first, should really have been expected. First, the range of temperature variations deduced from the measured intensities of thermal emission in the region of the radio-frequencies turned out to be *smaller* than those measured in the near infrared—the smaller, in fact, the longer the wavelength. Secondly, the maxima and minima of the emission in the domain of microwaves did not follow in phase the altitude of the Sun above or below the horizon, but *lagged* behind their surface (i.e. infrared) values by amounts increasing again with the wavelength. At 1–2 mm wavelengths, the lunar temperatures were found to vary between −100 °C and +30 °C with a phase-lag of 2–3 hours. At centimetre wavelengths ($\lambda = 1\cdot25$ cm) this range diminished to −70°C to −30°C with a time-lag of more than three days, while at metre wavelengths a mean temperature of around −30 °C remained constant day and night.

What can be the cause of these peculiar phenomena? The basic explanation is that *thermal emission at different frequencies does not originate at the same depth* but arises from layers which are, in general, the deeper, the longer the wavelength. In other words, the lunar surface, which is opaque to visible or infrared light, becomes partially transparent to microwaves; the lower their frequency, the deeper we can penetrate with their aid. But if the effects of the 'diurnal heat wave', caused by insolation, are damped as rapidly, and continue to lag in phase as much as is indicated by radio observations, the only reason for this must be the very poor thermal conductivity of the lunar surface which is characterized by a coefficient of such conductivity far smaller than that of any solid substance known on the Earth. Therefore, by and large the exposed surface of the Moon cannot consist of solid rocks, but must be covered by pulverized material in which heat can flow only through the corners where individual dust grains or pebbles happen to come into actual contact. In other words, it is mainly the reduced effective cross section of heat flow, brought about by the gradual pulverization of the lunar surface produced by the mechanical action of external events, that renders this surface such a good insulator.

A limit to this kind of prospecting in depth at a distance is imposed by the attenuation of lunar thermal emission at long wavelengths. In fact, at metre wavelengths it can scarcely be distinguished from instrumental noise (or sky background). However, if we illuminate the Moon from the Earth by *radar* pulses corresponding to decametre (10–20 m) wavelengths, observable echoes can be obtained from depths at many dozens of metres. The intensity of these echoes indicates that the surface of the Moon is highly fractured and consists of loose rubble down to many times these depths.

Conversely, the existence of any local variations of surface temperature can be used to study the different degree of compactness in different places on the Moon, especially from observations made during total eclipses of the Moon when the 'cold wave', which sweeps over our satellite when it enters the shadow of the Earth, makes its surface temperature fall and rise again at a truly frantic rate. As we have already mentioned, the topmost layers of the surface then cool off by 200 °C or more within a few hours because the heat capacity of its fine soil is so small. However, large stones or boulders possess a much greater capacity to absorb heat during the day and to emit it more slowly in the course of the night. During an eclipse which lasts only a few hours, these boulders will hardly cool off at all and will continue to emit heat that warms up their surroundings. In effect, a group of them then act as 'heaters' of the surrounding landscape, and an observer with eyes sensitive to their thermal radiation in the infrared would see the landscape dotted with 'hot spots' wherever boulders abound. Such spots were detected by the hundreds by R W Shorthill and J Saari during the total eclipse of the Moon visible in Egypt on 19 December 1964 (see plate 20).

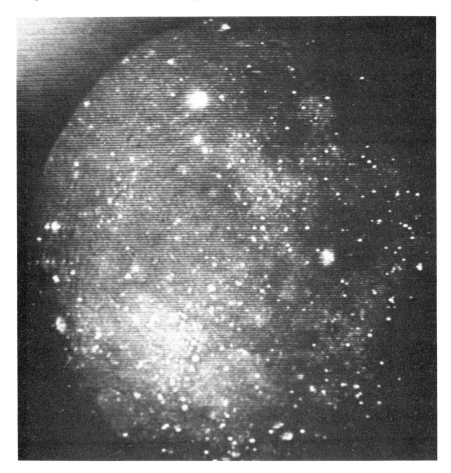

Plate 20. Not a picture of the starry sky at night, but of the face of the eclipsed Moon scanned in its thermal radiation! This is what we should see during eclipses if our eyes were sensitive to deep infrared light (around 10 micrometres effective wavelength). The picture above was reconstituted from infrared scans of the lunar face by R W Shorthill and J Saari with the 74 inch reflector of Helwan Observatory at Kottamia in Egypt during the total eclipse of the Moon on 19 December 1964. Each bright spot of the image corresponds to a region of enhanced thermal emission, indicative of less rapid cooling of the ground in the course of the eclipse. The three most conspicuous regions coincide in position with the craters Tycho (top) and Copernicus (right).

Subsequent photography from the Orbiters of 1966–1967 confirmed that the location of these spots corresponded to regions where large boulders did indeed abound on the surface.

Structure and Composition of the Lunar Surface

With the advent of the manned missions in 1969, the previously held views on the nature of the lunar surface could be tested directly. The outcome of the reconnaissance and experiments performed *in situ* broadly confirmed most expectations. In particular, the measured local temperatures agreed closely with those inferred from the measurements of thermal emission made previously at a distance. The fine structure of the lunar surface was also found to be consistent with that inferred previously from the 'photometric function' of changes in lunar light during the course of the month.

The closer acquaintance with the structure of lunar rocks also allowed us to extend our knowledge of the cratering processes on the Moon to much smaller formations than those known to us before 1969. While telescopic observations from the Earth enabled us to discern craters down to about 1 km in size, and the soft-landers of 1966–1967 disclosed details of the surface to within 1 mm in the immediate proximity of the spacecraft, lunar rocks brought back by the Apollo missions showed that their surfaces were 'etched' by impacts of solid grains of interplanetary dust (cf last section of Chapter 6). Such grains, only micrometres in size and $10^{-8}-10^{-14}$ g in weight, are being swept up in untold numbers by the Moon, as well as by the Earth on their journey around the Sun. On the Earth they fall as slowly through our atmosphere as through a sieve and eventually (after weeks or months) gently touch the ground to disappear into oblivion. However, on the Moon, where there is no atmosphere, they impinge on the surface with a cosmic velocity of several kilometres a second and on impact produce microcraters 1–1000 μm in diameter, examples of which can be seen in plate 21.

The number of grains in space capable of producing such craters, and intercepted by the Moon in the course of time, is truly enormous. According to current estimates, each square millimetre of the lunar surface receives a direct hit by a particle weighing 10^{-6} g or less once every 10^6 years, which means that several thousand of them should have scored hits since the formation of the lunar crust. This sets the timescale on which the microrelief of the lunar surface can be formed, destroyed, and recreated by the impact of micrometeorites raining down upon it incessantly, just as raindrops on Earth may checker a dusty surface with their pockmarks. It is primarily this kind of 'etching' which enables the microstructure of the lunar surface to backscatter light. The fact that this property is shared by all types of lunar ground, both continents and maria, provides a powerful argument in support of our

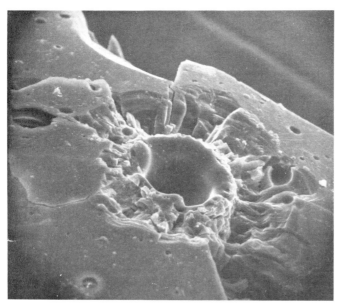
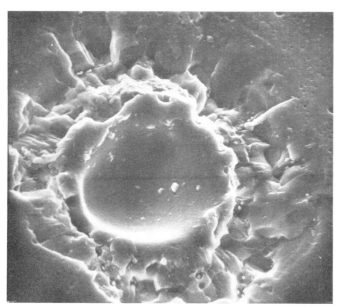

Plate 21. Examples of impact-produced microcraters in the lunar rocks brought back by the Apollo missions. The size of each field is approximately 10 micrometres across and the central pits (a few micrometres in diameter) show evidence of melting by the heat derived from the kinetic energy of impact. The microphotographs were taken with a magnification of 2250 by J L Carter of the University of Texas.

view, for what else but an external influence could impress the same uniform microrelief all over the Moon?

According to the latest estimates (D E Gault 1974) the topmost half millimetre of the lunar surface is 'ploughed over' by micrometeoritic action once in about 10^4 years, while 10^6 years would be sufficient to do so to a depth of 3 mm, and 10^8 years to 10–100 cm. Only the topmost 0·1–1·0 mm veneer is subjected to intensive churning at any one time, and it is this layer which is responsible for most of the optical properties (reflectivity, polarization) of visible moonlight. It goes without saying that this continuous action of micrometeoritic infall is occasionally supplemented by a much more intensive downpour of lunar material ejected by each major crater-forming impact. An ejection of material capable of giving rise to craters like Copernicus (plate 12), Theophilus (plate 13) or Tycho (plate 14) would cover the entire globe of the Moon with a layer of debris 10–100 cm thick, while impacts which gave rise to Mare Imbrium or Mare Orientale could (and apparently did!) bury the pre-existing landscape with secondary ejecta down to a depth of many metres. Such impacts, however, represented isolated events, while, in between these events, micrometeoritic infall continued to grind the lunar surface very slowly, but exceedingly finely.

Another corroborative point should be stressed here: although the lunar backscattering properties appear to be duplicated on Mercury (Chapter 4), to the same extent this is not true of Mars (Chapter 5). The difference in surface structure which this suggests may arise from the fact that whereas the Moon or Mercury do not possess any atmosphere to speak of (so that micrometeorites can impinge on their surfaces with undiminished cosmic velocities), the Martian surface is partly protected by an atmosphere which, albeit tenuous, can moderate the fall of such particles. The properties of the Martian atmosphere will be discussed in Chapter 5, but here we would like to examine just how much gas there is around the Moon.

All optical tests carried out from the Earth to identify the presence of gas around the Moon were negative and could only impose an upper limit to its hypothetical pressure and density. It was not until the advent of mooncraft and, in particular, of the last three Apollo missions that the amount of free gas around the Moon could at last be established by direct measurements. The outcome revealed its amount to be so small that we cannot speak of any atmosphere at all, only of an *exosphere* in which individual particles (atoms or ions) describe essentially free-flight trajectories in the prevalent gravitational (or electrostatic) field. Each planetary atmosphere is bound to peter out into such an exosphere on its outer fringe where it borders on interplanetary space, but on the Moon this exosphere reaches down to the solid surface itself. It is this surface which effectively controls the 'temperature' of the exosphere, the particles of which possess a mean free path of several hundred to a few thousand kilometres and a mean time of free flight of the order of several hundred seconds. Eventually, such particles—if neutral—will be ionized by solar radiation and escape from the

Moon by spiralling along the magnetic lines of force carried by the solar wind, unless they collide with the lunar surface and become neutralized again by electron capture.

The lightest constituent of the exosphere is, of course, hydrogen, and its principal external source is the solar wind. As is well known, the sunlit side of the Moon is continuously bombarded by the Sun's corpuscular radiation whose predominant constituents are protons. Together with the accompanying electrons of a neutral plasma, these travel through space usually with velocities between 300 and 400 km s^{-1} and possess kinetic energies of about 400–500 eV. Before impact on the lunar surface the protons are neutralized by electrons evaporating from it under the ionizing effect of solar ultraviolet radiation, so that what actually strikes the rocks of the lunar surface is really hot hydrogen gas.

Now the noon-time temperature of this surface is about 400 K, and a recoil from such a surface would endow its atoms with an average velocity of 2·9 km s^{-1}, which is sufficient for escape from the gravitational field of the Moon. Only cooler atoms with lower velocities can be retained to constitute an exosphere of some 3000 km scale-height, comparable with the diameter of our satellite. On the other hand, xenon (which on account of its high atomic weight would only be endowed with a mean velocity of 0·23 km s^{-1} by collisions with the surface at 400 K) could constitute a much more compact exosphere with a scale-height of only 20 km.

How much of the hydrogen is actually being retained in the lunar environment? Spectroscopic measurements by W G Fastie and his co-workers aboard Apollo 17 have disclosed that only about 10% of the solar-wind protons are arrested by the Moon to provide for its exosphere. The total amount of neutral hydrogen does not seem to exceed about 1 kg of free gas at any particular time. The total amount of xenon, which escapes less readily, may be larger, but still not in excess of 1 ton. These amounts are certainly very small in comparison with those released in the lunar environment by human action. Between 1966 and 1967, the soft-landers and orbiters listed in tables 2 and 3 expended several tons of gases in connection with their astronautical manoeuvres in the proximity of the Moon. Each one of the seven Apollo missions between 1969 and 1972 added more than 20 tons of exhaust gases, mostly of a sufficiently high molecular weight as to make their dispersal from the gravitational field of the Moon a very slow process. Altogether more than 160 tons of gases were thus released in the lunar environment within six years, but the suprathermal ion detectors of the successive ASLEP packages (see plate 9) failed to detect their presence. Why are they not there? The likeliest answer is that these gases must have been very rapidly *absorbed* by the material of the lunar crust and thus bound in solid state. Moreover, the outcome of all experiments carried out so far discloses that natural degassing of the lunar interior must now be virtually at a standstill—an outcome well in accord with what we have learned by seismic and other methods about the rigidity and low temperature of most of the lunar globe.

Because of the extreme tenuity of the lunar exosphere reaching down to the surface, the latter is continuously exposed to irradiation not only by sunlight of all wavelengths (from infrared to x-rays), but also by the entire corpuscular output of the Sun as well as of other objects in the Galaxy. The solar output consists mainly of proton particles of energies less than 500 eV for the solar wind produced by the 'quiet-Sun' and of cosmic-ray particles of energies 10–100 MeV produced by disturbances on the Sun. Even higher energies are produced by certain sources within the Galaxy. Unlike the protons of the solar wind, which are intercepted by the Moon and converted into neutral hydrogen, cosmic-ray particles lose energy mostly by ionization. Sometimes, however, particularly energetic collisions give rise to nuclear reactions producing radioactive nuclides with half-lives of the order of 10–100 million years, times very much shorter than the age of the Moon itself. Many radio-isotopes of this kind are known (e.g. ^{22}Na, ^{26}Al, etc) and their presence in different layers of lunar rocks can be ascertained by microchemical methods.

In addition, the impacts of these particles can impress latent tracks in crystalline rocks. Although of microscopic dimensions, the tracks can be made visible by coating the rock with suitable chemicals (such as silver nitrate) and an example of a lunar rock treated in this manner is shown in plate 22. Such tracks can indeed be regarded as 'microcraters' caused by very high velocity impacts of heavy elementary particles and atomic nuclei accelerated sometimes to speeds slightly less than that of light. The depths of the tunnels left behind in crystalline rocks can be regarded as a measure of the energy of the impinging particles.

Before 1969 such tracks were known only from the study of meteorites, but the lunar rocks brought back by the Apollo missions showed similar effects of corpuscular irradiation. A concentration of the artificial nuclides produced by this irradiation can indicate the time during which the respective rock was exposed to (i.e. within reach of) the cosmic rays capable of bringing about the particular transmutations. An asymmetry in distribution of the respective spallation products† over the surface rocks can also disclose the extent to which these rocks 'wiggled' in their positions, or were overturned during this time.

The 'exposure ages' of lunar rocks gathered in different localities (as distinct from the radiometric ages of solidification) turned out to range between 20 and 200 million years. Mechanical disturbances associated with impacts or other upheavals, such as landslides, may either bury rocks to sub-surface depths outside the range of penetration of most cosmic rays, or again exhume them from these depths to be closer to the surface. A top layer about one metre deep seems to be ploughed over by all kinds of erosion processes in a time interval of the order of 100 million years, approximately the time between the Cretaceous period on the Earth and the present.

†Spallation products are new nucleons generated by the impact of cosmic rays.

The Moon: Our Nearest Celestial Neighbour

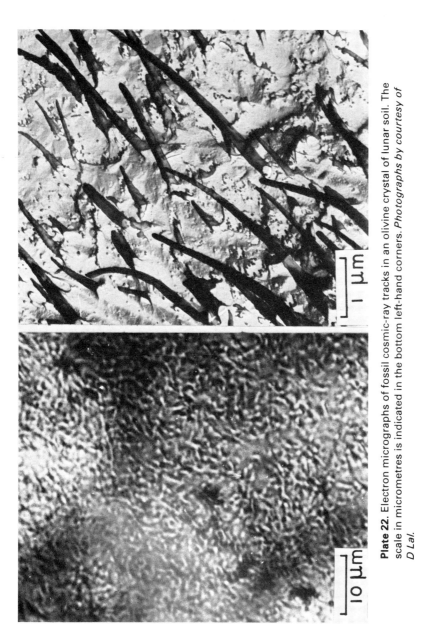

Plate 22. Electron micrographs of fossil cosmic-ray tracks in an olivine crystal of lunar soil. The scale in micrometres is indicated in the bottom left-hand corners. *Photographs by courtesy of D Lal.*

The extreme slowness of any changes on the Moon gives an impression of utter desolation, as may be seen from the lunar landscape shown in colour plate 3. The lunar plains are indeed more barren than any terrestrial deserts, and pitch-black lunar caves are abodes of eternal silence. A spider's web stretched across the dim recess of any cavity would have a good chance of remaining undisturbed for millions of years. Indeed, almost nothing ever happens on the Moon, locally or globally (apart from the host of man-made mooncraft which have disturbed its peace in the last 20 years!). The changes prompted by the processes discussed earlier in this section continue to grind the Moon's face very fine, but at an exceedingly slow rate. Not a very exciting place for holiday-makers, perhaps, but for the scientist an Aladdin's cave full of heavenly wonders, both on the ground and in the sky!

Chronology of the Lunar Surface

In the preceding sections we discussed the nature of the lunar surface and described the principal types of formations encountered on its landscape. The aim of the present section will be to consider the *collective* aspects of the distribution of such formations over different types of lunar ground in order to disentangle their stratigraphy and to calibrate it with an absolute time-scale.

As we stressed before, all features visible on the lunar surface must originate from either the internal processes for which this surface represents merely the 'boundary condition', or from external impacts of the particulate contents of space ranging from interplanetary dust and micrometeorites to meteorites, asteroids and comets. For these the lunar surface merely constitutes an 'impact counter'. The scars left behind by such impacts do not fall prey to any kind of erosion (other than seismic) which could gradually obliterate old features; pre-existing formations can be effectively destroyed only by fresh impacts. If so, however, then the cumulative effects of impacts may conceal a *stratigraphic time-sequence* reflecting the events of bygone days for which we no longer possess any fossil record on the Earth because of the greater internal activity of our planet.

The adoption of such an attitude does not imply that we regard all formations on the lunar surface to be of external origin. Far from it, for internal causes must have been operative to produce formations such as domes, rilles, or wrinkle ridges which cannot be accounted for by the direct intervention of external agents. Nevertheless, from the large preponderance of brecciated rocks on the lunar surface we are convinced that most crater-like formations on the Moon—and in particular the largest among them—are of external origin. As we shall see, many of them belong among the oldest landmarks still visible on the surface of the Moon; the problem is only to arrange them in a consecutive time-sequence.

There are, indeed, three different and independent (though

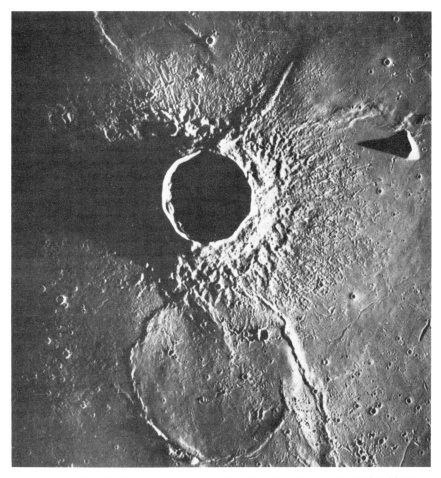

Plate 23. A photograph of a typical 'ghost' crater (Lambert R) in Mare Imbrium south of the crater Lambert (above) taken by the metric camera of the Apollo 15 mission in July 1971 from an altitude of 98 kilometres above the lunar surface. Another ghost crater (Daguerre) in Mare Nectaris east of the crater Theophilus is shown in plate 27. *Photograph by courtesy of NASA.*

largely qualitative) ways which can be invoked to this end: (i) the principle of overlap; (ii) the degree of ruggedness; and (iii) the ground reflectivity. Of these three, the principle of overlap is perhaps the most obvious and dependable one. If two craters overlap each other, we conclude that the one with an unbroken rim must be more recent than the crater whose rim was damaged or entirely removed. For the case of two overlapping craters of comparable dimensions, such as the Theophilus and Cyrillus pair in plate 13, Theophilus should, accordingly, be of more recent origin than Cyrillus. Similarly, a large crater containing smaller ones on its floor, such as Clavius for example (see plate 11), must obviously be older than any formations seen on its floor, for the process that raised the ramparts would not have left them undisturbed. In the same way, all fresh-looking craters on the floors of the maria must be younger than the respective mare itself; those antedating its formation at best protrude through its lava cover with the rims of their ramparts to give the appearance of a 'ghost' crater (see plate 23).

The second criterion, that of the ruggedness of the terrain, is partly a consequence of (i). If we accept the view that most of the stony sculpture of the lunar surface is due to external impacts, there is no escape from the conclusion that the oldest parts of this surface are those which are the most rugged. Irrespective of the rate at which the Moon intercepts intruders from interplanetary space, the oldest parts of its surface should obviously have accumulated the greatest number of scars. If this is indeed the case—and there is scarcely any doubt about such a conclusion—then the oldest regions on the Moon occur predominantly on its far side or, in the visible hemisphere, around the south pole. Many parts of those areas seem indeed to have received a 'saturation' bombardment from space, implying that their oldest sculpture may have been destroyed completely (and possibly more than once) by subsequent impacts. Conversely, the smooth plains of the lunar maria should be of a much more recent date.

The third criterion, the degree of ground reflectivity, can be applied only to a relatively recent past. It is based on a well founded contention that all physical and chemical processes (sputtering, radiation damage, interaction with the solar wind) operative on the lunar surface tend to *darken* its material—the more so, the longer it remains exposed to such influences. We have already pointed out that the formation of primary impact craters on the Moon is bound to be accompanied by an ejection of a large amount of sub-surface material with velocities which, in the feeble gravitational field of the Moon, may carry its particles to distances of several hundred to a few thousand kilometres. Since the ejected material was not previously subjected to darkening influences, it will stand out conspicuously on a darker background in the form of bright aprons surrounding a young crater, sometimes spreading out tentacles of bright rays to considerable distances. The craters Copernicus (see plates 10 and 24) and Tycho (plates 10 and 25)—well known to telescopic observers of the Moon—can be taken as typical examples of impact formations whose ejecta have not yet been darkened by age to the reflectivity of the underlying substrate. A few smaller ones, such as the crater Aristarchus in Oceanus Procellarum or Giordano Bruno (plate 26), are still brighter. On the other hand, the rays of Theophilus are already very faint (plate 27), while Eratosthenes, for example, exhibits all the other structural features of these craters yet completely lacks rays.

It is more than probable that the diminishing brightness of the rays reflects the increasing age of the respective formation. If we knew the rate at which these ejecta fade away in the course of time, the brightness of the rays could be used as a measure of the time which has elapsed since the

Plate 24. A system of bright rays in Oceanus Procellarum surrounding the crater Copernicus. A similar bright apron of ejecta also surrounds the craters Kepler (to the west of Copernicus) near the right margin of the field, and Aristarchus (near the lower right-hand corner). An older impact crater Eratosthenes (to the left of Copernicus) lacks rays altogether. The photograph was taken with the 100 inch reflector of the Mount Wilson Observatory.

impact which gave rise to them. Similarly, if we knew the rate at which the surface of the Moon has been bombarded by external intruders in the past, the density of impact craters per unit area and its comparison with the effects produced by the same flux on the Earth (where individual events can be dated by geological methods) could again be used to estimate its age. Various estimates of such ages in the pre-Apollo days differed greatly from each other, but were unanimous in indicating that the lunar landscape as we see it today must have originated before Palaeozoic times on the Earth.

The real breakthrough in lunar chronology occurred in 1969 with the advent of re-entrant mooncraft, which enabled us to provide an

Plate 25. A photograph of the crater Tycho (see also plate 14) and its neighbourhood taken with the 120 inch reflector of Lick Observatory at Mount Hamilton. Note the dark halo of 'ballistic shadow' surrounding the central crater and overflown by the bright ejecta.

absolute calibration for the relative age of the lunar surface. As was mentioned in Chaper 2, the Apollo missions of 1969–1972 brought back some 382 kg of lunar rocks of different types (and collected in different localities) whose absolute ages could be established by radiometric methods. It has been known since the days of Rutherford that the ages of solid rocks can be ascertained from the progress of spontaneous disintegration of certain long-lived radioactive elements which are present in the rocks in small but measurable amounts. To mention only a few of these little 'radioactive clocks', ^{238}U and ^{235}U decay spontaneously into radiogenic isotopes of lead (^{206}Pb and ^{207}Pb) at an exponential rate so slow that one-half of the original substance is transformed into its resulting products in 4·51 and 0·713 billion years, respectively; whiie ^{232}Th decays into ^{208}Pb even more slowly, its half-life being 13·9 billion years. These transmutations pass through many intermediary stages, but their end-product of radiogenic lead is a solid substance which differs from 'ordinary' lead by its atomic weight and can therefore be isolated from the mixture by known methods. On the other

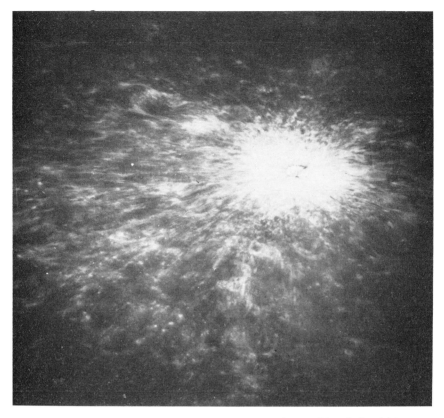

Plate 26. The lunar crater Giordano Bruno (on the Moon's far side) as photographed by the Apollo 8 mission on 24 December 1968. The white apron of bright rays suggests a relatively recent origin. *Photograph by courtesy of NASA.*

hand, ^{87}Rb decays into ^{87}Sr, or ^{40}K into ^{40}Ar, by a simpler transmutation involving the loss of only one electron ('beta-decay'). While the half-life of the rubidium–strontium metamorphosis is extraordinarily long (47·1 billion years), that of the potassium–argon pair amounts to only 1·27 billion years. This last decay also differs from all those mentioned before by the fact that its end-product is a gas and not a solid.

We may ask with incredulity: how is it possible to measure in the laboratory half-lives that are so long? The answer is implied in the enormous number (about 10^{22}) of atoms potentially available for such a transformation in any gram of matter which we can handle, a number larger in fact than the number of seconds in 10^{15} years and much larger than the number of stars in the entire Universe. While each individual atom of (say) ^{87}Rb may have to await its chance to emit an electron from its nucleus to transmute into an atom of strontium for a time longer than the age of the Universe, we do not

Plate 27. The craters Theophilus, Cyrillus and Catharina on the western shores of Mare Nubium (to the left). Some rays of Theophilus can still be dimly seen on the darker mare ground. Photographed with the 24 inch refractor of the Observatoire du Pic-du-Midi (Manchester Lunar Programme).

know the moment when this event may occur. It is a lottery; and so large is the number of players that although one-half of them will have to wait more than 47 billion years for a win, a very large number will draw winning tickets each second and tell us of their success by instant messages which our instruments can register. This kind of lottery is characterized by one rule very useful to the student of the subject: whether it is played in the laboratory or in celestial bodies of planetary size, the rate at which radioactive elements continue to disintegrate is strictly constant and wholly oblivious to the surroundings. It is only in bodies of stellar, rather than planetary, size that much higher pressures and temperatures begin to exert an influence on the speed of the game, but on the Moon or the planets this is not the case.

On the Moon and on other planetary bodies, the 'radioactive clocks' represented by an ensemble of unstable atomic nuclei possess another feature in common: they start to mark time only from the moment when the material containing them has solidified. The dials of these clocks indicate the proportion in the sample of the daughter product to the mother substance and are automatically reset to zero whenever the respective rock has been remelted (because the mother and daughter products could then be separated by chemical processes such as diffusion or volatilization). The situation is analogous to the well known ^{14}C method used by archaeologists to date the age of any organic tissue which is of interest to them. Here also the age is measured from the time when the organism stopped its intake of cosmic-ray-produced ^{14}C from the atmosphere at the time of its death.

When methods of this type were applied to lunar rocks to establish the time which has elapsed since their solidification, the results were surprising, so much so that it has taken several years of hard work to unravel their story. Although further modifications are not ruled out by an acquisition of new material in the future—for after all the amount of lunar rocks (not all of which were analysed by radiometric methods) selected at random constitutes only a very small sample of the lunar mass—we are confident now that it bears out the essential features of this story.

The first great surprise awaiting us in 1969 was the extraordinary *old age* of the lunar rocks: none brought back then (or since) has proved to be less than 3 billion years old, and the majority were considerably older. It soon became apparent that the period of rock-forming activity on the Moon antedated by and large the known geological history of the solid crust of the Earth; the oldest known rocks on the Earth and the youngest on the Moon being of a comparable age. The oldest rocks brought back from the Moon turned out to be between 4·5 and 4·6 billion years old, which is, very approximately, also the age of the oldest known chondritic meteorites (carbonaceous chondrites) intercepted by the Earth. The coincidence between these two ages leads us to conclude that *the time-span of 4·6 billion years signifies the age of the entire solar system with all its constituents*.

The oldest samples of lunar rocks have not been preserved in large lumps. In point of fact, most particles of that age brought back from all the landing places are described as 'fines' or 'soil'. The reason why next to no material of that vintage came down to us in larger pieces is no doubt the fact that the subsequent history of such matter was a very turbulent one. We have many reasons to believe that, for the first few hundreds—nay, tens—of millions of years, the lunar surface was subjected to so intense a bombardment by cosmic debris from surrounding space that most of the larger rocks did not survive in one piece, but were gradually shattered into the finer debris which covers the Moon today. Since the oldest rocks were not only shattered in the course of this initial bombardment, but were also transported all over the Moon as impact ejecta, it is virtually impossible to assign any specific locality of origin to such 'fines'. The oldest sizable lunar rocks now in our hands are not much more than 4·3 billion years old, which means that these solidified when the Moon was already about 300 million years old. However, a gradual intermingling of such rocks by the mechanical action of further impacts continued after that date, so that it is difficult to assign any particular piece to any particular event.

Only the largest impact formations on the lunar surface—such as the basins which, if filled with basaltic magma, are known as the lunar maria—threw out a sufficient amount of ejecta to make possible the identification of the material produced by such events. The task of establishing the absolute dates for the origin of the lunar maria can, in principle, be approached in two ways: from radiometric dating of brecciated rocks which, presumably, acquired their texture by the impact that excavated the respective basin; and from the similar ages of the basalts which filled the depressions. This has indeed been done for the majority of conspicuous mare basins on the near side of the Moon (mostly visible to the naked eye) and the results are summarized in the table below.

Radiometric ages of Mare-forming events

Mare	Age (in 10^9 years)	
	Breccias	Basalts
Serenitatis	4·26	3·79±0·04
Nectaris	4·25±0·05	
Foecunditatis	4·20±0·04	3·5
Tranquillitatis	4·20±0·05	3·7±0·2
Humorum	4·16±0·04	
Crisium	4·05±0·05	
Imbrium	3·95±0·05	3·3±0·1
Orientale	3·85±0·05	

The data in the table disclose the following important results on the past history of the Moon:

(i) As indicated by the radiometric ages of the associated breccias, the principal maria on the near side of the Moon were excavated 3·8–4·2 billion years ago (and 400–800 million years after the formation of our satellite) by isolated impacts spread over an interval of some 400 million years. The impact which gave rise to Mare Serenitatis was probably the first of a discrete series of events which disfigured the face of the Moon in this manner. Mare Orientale (see plate 17) represents the last and best-preserved addition of its kind to lunar topography. The impact which produced Mare Imbrium (plate 16)—the best-known formation of this type directly facing us—must have occurred when the Moon was just over 600 million years old, while the dating of Oceanus Procellarum—the largest mare formation on the Moon—is still uncertain.

(ii) As indicated by the radiometric ages of mare basalts, the *flooding* of the maria by basaltic magma occurred between 3·3 and 3·8 billion years ago, that is, some 400–700 million years *after* their basins had been excavated. The two events which gave rise to the lunar maria, namely the excavation of their basins by impacts and their subsequent flooding, did not, therefore, follow each other immediately but were actually separated by hundreds of millions of years. No mare basalts appeared on the lunar surface in the first 800 million years of its existence, nor were any added more than 600 million years later. If we could have had a look at the Moon (say) 4 billion years ago, its face would not have shown any 'dark spots'; while about

3·2 billion years ago, to the naked eye its face would have looked very much the same as it does today.

The conspicuous time-gap between excavation of the lunar maria and their subsequent flooding by lava gives rise to several questions which demand a satisfactory answer. What happened to the heat which must have been produced by the impacts that excavated the maria; and what was the source of the heat needed to produce the magma which subsequently flooded the maria? We know from the partition of energy in impact phenomena that about one-half of the total kinetic energy of the intruder destroyed by impact should have reappeared in the form of heat. If this heat did not spend itself in the melting of the lava, what else did it do? In a preceding section we mentioned that an impact crater 900–1200 km across (of the dimensions of Mare Orientale or Mare Imbrium) should also have been more than 200 km deep, while in reality the surfaces of these maria do not deviate from a mean sphere of 1738 km radius by more than a few kilometres (see figures 1 and 2). In addition, we have stressed that under ordinary circumstances the outer layers of the Moon down to this depth are much too rigid to admit a spontaneous return to isostasy† even in a time as long as 4 billion years. Was it the heat suddenly liberated by an impact of this magnitude which softened the rigid crust and facilitated a return to hydrostatic equilibrium that reduced the original depth of the 'dry' basins before they were overflowed by magma at a later time?

The physical structure and chemical composition of the basaltic magma brought back from the marial regions disclose that the source of the original magma is to be found at least 400 km below the Moon's surface, where 'volcanic pockets' of molten lava could have been formed by cosmochemical processes of an as yet unspecified, but local nature. The fact that such pockets of underground lava coincided in position with the basins excavated by overhead impacts—at least on the near side of the Moon—suggests that the delayed stresses produced by impacts had something to do with their formation.

On the other hand, the amount of lava which actually reached the surface was probably quite small. Several reasons lead us to this conclusion. One is the existence of 'ghost' craters (for an example see plate 23), pre-existing on the floor of the original basins and with ramparts still protruding through the flood that later engulfed them. The existence of these formations in many maria provides, by itself, an indirect but sufficient proof that the excavation of the basins and their subsequent flooding with lava had to be separated by a considerable time. Since the pre-existing sculpture would have been totally destroyed by the impact which scooped out the basin, a considerable time had to elapse before the subsequent impacts of smaller-calibre missiles could have again cratered the basin in its 'dry' state by a new population of such formations before these were eventually flooded.

†The maintenance of hydrostatic equilibrium.

Smaller craters formed in this way were bound to be flooded completely; only those high enough can still protrude from the present mare floors. As, moreover, the heights of these ghost craters can be estimated by Ebert's rule from their observed dimensions, the smallest formations still visible on the surface can indicate to us the depth of subsequent flooding. A determination of the depth of flooding disclosed that basaltic lavas cover most of the lunar maria only to a depth of a few hundred metres. The average thickness of the overlay of Mare Tranquillitatis appears to be only 500–600 m, with a maximum depth not exceeding twice this amount. Mare Imbrium is no deeper; only Mare Nectaris (the oldest of the walled basins and whose walls are still partly preserved) attains an average depth of some 900 m and a maximum depth close to 1500 m. In all cases, the basaltic overlay of the mare plains appears to be only skin deep and does not exert an appreciable load over their substrate.

The fact that basaltic lavas seem to be spread out over the underlying substrate only in the form of a thin veneer whose thickness is very small in comparison with its distension is borne out by other topographic features of the lunar maria discovered recently by laser altimetry. Had the lunar basins been filled at any time with molten magma down to a considerable depth, the surface in which its material would have solidified should have been an equipotential (level) surface, the form of which would be everywhere normal to the direction of local gravity. The actual measurements of vertical profiles of the mare ground overflown by the Apollo 15–17 laser altimeters (see figures 1 and 2) disclosed that this was not the case. Some maria, in particular Serenitatis or Imbrium, appear to be noticeably *inclined* to the horizontal (both in the same sense), and the western shores of the Imbrium–Procellarum complex are elevated over its eastern shores by more than one kilometre. Such a situation can be understood only if the form of the present surface is governed by the slope of its solid floor overlain by a relatively thin veneer of basaltic material, rather than representing a free boundary of frozen fluid.

A third phenomenon indicative of the shallow depth of the lunar maria was mentioned in the section dealing with seismic evidence. On p 48 we stressed that the principal feature of seismic records distinguishing moonquakes from earthquakes is the very long duration of their 'echoes', a duration ascribed to the scattering of seismic waves in a highly fractured surface layer about 20 km deep. If the lava cover were to extend to this depth, it could no longer be fragmented to the requisite extent, for by the time lava reached the surface (3·2–3·8 billion years ago) the principal epoch of celestial bombardment causing fragmentation of the surface was already over. This is attested to not only by low crater counts on solidified mare ground, but also by the absence of such a fractionation of old terrestrial rock strata which should have absorbed an equal punishment by celestial projectiles per unit area. The only logical explanation for the observed phenomena seems again to be offered by the hypothesis that the fractured rock layer which produced the scattering of seismic waves was overlain by a thin veneer of lava 1–2 km deep. This veneer played the role of an elastic boundary ('waveguide') to the underlying layer, some 20 km deep, and was highly fractured by the impact which excavated the original basin.

Origin and History of the Moon

One of the principal facts which have emerged from the radiometric studies of rocks brought back from the lunar surface by different space missions is the age at which the Moon originated as a solid body. It can now be dated with relative precision to 4·6 billion years (with an uncertainty of only about one per cent). This, at least, is the age of the oldest solid particles ('fines') which occur widely over the surface of the Moon and which have not been remelted since. Moreover, this age agrees closely with that of the oldest (chondritic) meteorites, and is generally regarded to date the origin of the solar system as a whole. The creation of the Moon represented only an isolated episode of this event.

Unfortunately, the actual details of this creation still continue to elude us. That the Moon originated by a rotational fission of the Earth is, while not dynamically impossible, unlikely on many other grounds. A somewhat less unlikely possibility is that the Moon was captured by the Earth after its formation. If so, this would imply that the Moon was captured prior to the main crater-forming period on the lunar continental land masses, for it is most unlikely that their pre-existing sculpture would have survived so brutal an experience. The capture of a celestial passer-by into a closed orbit would have required a dissipation (through inelastic tides) of so much energy into heat as could be raised only at a very close approach. Such tides would have obliterated without doubt most of the pre-existing surface structure of the Moon, as well as of the Earth. We have, of course, no evidence whether or not the Earth itself experienced so gruesome an ordeal, for no terrestrial rocks of that age survive intact to this day. A history of the stony strata of the Moon (which reaches back much further in time) also contains, however, no recollection of such an event. If it took place, all we see and find on the Moon today must have been formed since that time.

Chemical reasons lead us to conclude that the Earth and the Moon originated in the same general region of the solar system. For example, in lunar samples the isotopic composition of oxygen—the most abundant element of the lunar globe by weight—is indistinguishable from that of terrestrial oxygen, while the proportion of ^{16}O to ^{17}O and ^{18}O varies measurably in other parts of the solar system. On the other hand, in contrast with the chemical composition of the terrestrial crust, lunar rocks are systematically depleted in more volatile elements, a fact which leads us to conclude that the solid lunar material was never a part of the Earth or of any other more massive body.

The Moon is most likely to have originated from a low-temperature agglomeration of *solid* particles in the proximity of the Earth and in orbit around the centre of mass of the nascent Earth–Moon binary system. The smallness of the mass of the lunar end-product is sufficient to ensure that free gas could have played only a very subordinate role in this process. The mass of the Moon is too small to attract (or retain) any significant amount of gas or other more volatile elements; oxygen, for instance, constituting not less than 50% of the lunar globe by mass, must have been acquired in solid state in the form of pre-existing oxides. Unfortunately, none of the facts at our disposal can specify the distance from the Earth at which the creation took place. The distance was no doubt small in comparison with that which separates us from other nearby planets, and probably also small in comparison with the Earth–Moon separation at the present time, but not small enough for the Earth's gravitational attraction to interfere with the formative process which gave rise to the Moon.

The earliest extant parts of the lunar surface are—we repeat—old enough to have seen our Galaxy in its earlier stage of evolution and to have witnessed the Sun in its last throes of contraction towards the Main Sequence†. If lunar rocks could talk, they would no doubt tell a most interesting story. Some of them are actually in our hands but, unfortunately, their age and heavily damaged state make it very difficult for us to loosen their tongues and make them respond to more detailed laboratory inquiries.

Since the time of the origin of the Moon, its surface has passed through many vicissitudes. These can be divided into the following four main epochs:

(i) An intensive bombardment by particles and bodies left over in circumlunar space after the formation of the Moon. The omnidirectional nature of such impacts (discussed on p 62) leads us to conclude that, prior to impact, these particles were mainly in circumlunar orbits or were dynamical members of the Earth–Moon system. As a result, their impact velocities were generally low, of the order of 2–3 km s^{-1} relative to the lunar surface. The 'saturation bombardment' evidenced by large continental areas of the lunar surface (in particular, on the far side) probably goes back to the first 200–300 million years of the Moon's existence, and took place shortly after the formation of the lunar globe and before the supply of solid remnants was depleted by the gravitational attraction of the Earth as well as of the Moon. Impacts of particles in heliocentric orbits contributed to the havoc and mutilation of the lunar surface at that time, but probably only to a subordinate extent.

(ii) After most of the solid small fry in the Earth–Moon gravitational domain were swept up from space by collisions with the surfaces of these two bodies, between 400 and 800 million years of its existence the

†The Main Sequence of a star is the evolutionary stage where it derives light from a transmutation of hydrogen into helium.

Moon suffered a series of the greatest impacts of all—those which excavated the basins of the future maria. While smaller craters with diameters less than 200 km are distributed on the Moon essentially at random, larger basins show a distinct preference for the Moon's equatorial belt. This fact by itself does not rule out the possibility that the bodies which excavated the basins were in heliocentric orbits, for the lunar equator does not deviate very much from the ecliptic. However, the conspicuous clustering of such basins (at least of those which eventually became the maria) on the lunar hemisphere facing the Earth strongly suggests that the impinging bodies were not occasional visitors from interplanetary space revolving around the Sun, but were regular members of the Earth–Moon gravitational domain and even, quite possibly, satellites of the Moon itself. In other words, the meteorites or planetesimals whose impacts scooped out the basins for the future maria on the Moon's near side may have been lunar satellites of a size comparable to the Martian satellites Phobos and Deimos (see plates 45 and 46). Because of their larger masses, such bodies will always be the last to collide with their respective planets long after the small fry have been so obliterated. Mars has retained two up to the present time, and there is no reason why our Moon should not have retained a swarm of them for the first 400–800 million years of its existence. It should be borne in mind that low-velocity impacts of planetesimals only three to four times the size of Phobos or Deimos are sufficient to excavate basins on the Moon of the size of Mare Imbrium or Mare Orientale; even an impact by a body as small as Deimos would have scooped out beds for the future Mare Crisium or Mare Humorum. Before we became closer acquainted with Mars and its natural satellites by means of spacecraft, theories elaborated (by H C Urey and others) on the nature of the Imbrium collision may have carried an esoteric tinge, but this should no longer be the case today.

(iii) Some 400–500 million years after the formation of the basins, those on the Moon's near side were overlain by a thin cover of basaltic magma which gave the lunar maria their present appearances. The heat which melted this lava must have been of an internal origin, but the amount of lava carried to the surface could only be small. Moreover, the youngest sample of basaltic magma collected on the Moon so far is only 3·16 billion years old. From that time the Moon has become tectonically inert.

(iv) In the past three billion years—a time which represents more than two-thirds of the age of our satellite—nothing much has happened on the Moon, apart from occasional impacts by interplanetary intruders whose heliocentric orbits were on a collision course with that of the Moon. Such occurrences, which may produce impacts with higher velocities than the majority of the earlier events of this type and therefore leave a somewhat different 'signature' on the surface, are becoming increasingly scarce. After 4·6 billion years of continuous gravitational sweeping of interplanetary space by all the planets, not too many solid remnants are left.

Unprotected by any atmosphere, the Moon continues to expose its wrinkled face to all cosmic influences which may act upon it and continues to accumulate scars of new impacts, albeit at a diminishing rate. Those acquired in (say) the last billion years can be identified and dated by their systems of bright rays. The age of the crater Copernicus (plate 12) is known very well, for its ray material (as shown in plates 10 and 24) happened to splash onto the area of the future landing site of Apollo 12. The radiometric age of this material turned out to be very close to 900 million years. The conspicuous crater Tycho near the Moon's south pole (plates 14 and 25) may be of a comparable age, while Theophilus (plates 13 and 27) is older. Only craters as bright as Aristarchus in Oceanus Procellarum (or a few smaller formations still brighter) may have been added to the lunar surface since the time of the Palaeozoic Age on the Earth. It is possible that these were the last sizable events of this kind which the long-suffering face of the Moon had to endure. And, in the aeons of time to come, the surface of the Moon is likely to become more and more withdrawn in its petrified grimace reflecting the state of the solar system in days long gone and acting as a true fossil reminder of a past which no internal convulsion will disturb any more.

How different all this is from the Earth! The known history of our planet does not even reach as far back as that of the lunar 'fines'. In fact, the first billion years of its existence represent a veritable 'dark age' from which no geological evidence has come down to us. With the possible exception of certain rocks which were found recently in Greenland and seem nearly four billion years old, the terrestrial 'book of hours' only reads continuously from 3·5 billion years ago, an age not far from the time when the active part of lunar history was effectively over. However, since 1969 we have discovered on the Moon a source of material which has suddenly illuminated the earliest chapter in the history of the Earth–Moon system and which has enabled us to reconstruct an almost uninterrupted story of what has been going on in the inner precincts of the solar system since the time of its formation.

Unlike the Moon, our Earth continues to exhibit to space a changeable face of almost eternal youth: its face has been rejuvenated continuously by geological processes such as erosion; its land denuded by the joint action of air and water; its waters withdrawn from the oceans at the time of the ice ages; and, more importantly, its mantle has undergone continuous continental drift caused by convection and driven by heat from its interior. Very few parts of the present terrestrial continents or ocean floors are more than a few hundred million years old—less than 10% of the age of our planet. In contrast, the Moon, with its smaller mass and internal heat storage, can afford none of these means of cosmic cosmetics to make up her face. This face truly mirrors the passage of time and preserves a reflection of events which occurred long before our terrestrial continents were formed and long before the first manifestations of life flickered in our shallow waters. As a monument to the past, the Moon indeed constitutes one of the most important fossils of the solar system, and the correct interpretation of the hieroglyphs engraved by nature on its stony face holds a rich scientific prize.

So much for the Moon's past: but what of the prospects for its future? While the evolution of its surface has virtually come to a standstill, this is far from true for its kinematic properties or for those of the Earth–Moon system at large. On account of their isolation, these two bodies constitute a closed dynamical system whose behaviour must fulfil certain stringent conditions such as the *conservation of the total angular momentum* of their motions in the course of time. This total momentum consists of two parts: the rotational momenta of both components about their axes; and the orbital momenta of their motions around their common centre of mass.

If both partners of such a union were rigid (i.e. not susceptible to any deformation), or again perfectly deformable (i.e. behaving as a non-viscous fluid), then the rotational momenta of such bodies would be independent of their orbital momenta, and both would have to be preserved separately. In the case of deformable components capable of raising *tides* on each other this will not be true, as imperfect rigidity (or fluid viscosity) would cause tides to *lag* behind the instantaneous field of force. This asymmetry of deformation would then produce a torque to couple the rotational and orbital momenta through *tidal friction*. Now both the Earth and the Moon are imperfectly deformable bodies and, therefore, capable of a momentum transfer by such a mechanism. In the case of the Moon whose axial rotation is, and always has been, synchronized with the mean angular velocity of its orbit, any interchange between the rotational and orbital momenta can only be small and can only arise by virtue of the eccentricity of its relative orbit (giving rise to 'tidal breathing'). On the other hand, the Earth rotates today more than 27 times as fast as it revolves, while viscous tides (both bodily and oceanic) raised on it by the Moon tend to slow down its spin and transfer its rotational momentum lost by their action to the orbital momentum of the Earth–Moon system, a transfer which entails an increase in the distance between the two bodies.

The current level of tidal friction operative in this system indicates that the axial rotation of our planet should be slowing down at a rate corresponding to an increase in the length of our day by $1\frac{1}{2}$ milliseconds per century, a value consistent with the observed 'secular acceleration' of lunar and planetary motions and with changes in the length of the day measured directly by atomic clocks in the laboratory. A gradual increase in the length of the day over geological periods in the history of the Earth seems to be borne out also by biological evidence concerning the life cycle of corals. Certain ridges in the skeletal structure of these organisms have been identified with indications of their periodic diurnal, as well as annual growth. The corals of recent times exhibit some 365 diurnal growth rings per annum, but fossil corals from the Middle Devonian—a geological period occurring about 380 million years ago—exhibit about 400 rings per annum. Hence, the number of days per year in Devonian times should have been about 40 more than today, and if, as we have reason to believe, the length of the year in the meantime has remained essentially unchanged since Devonian times, this would imply a fractional increase in the length of the day by a factor equal to a ratio of 40 days in 380×10^6 years, that is, $2 \cdot 9 \times 10^{-10}$, a value in fair agreement with its astronomical determination by recent measurements.

Further palaeontological evidence of the diurnal and monthly effects on the growth of not only corals, but also of other fossil crustacea of the Palaeozoic Age indicates that in the Cambrian period (some 500 million years ago) a sidereal day on Earth had only about 21 hours of our present time (indicating a more rapid rotation of the planet), a year of 415 days, and 31·5 days in a synodic month. In the Devonian period (some 380 million years ago) the length of the day increased to 21·6 hours and that of the month decreased by one day. In the Upper Carboniferous (290 million years ago) the day increased to 22·6 hours and there were 30·1 days in the month. By the end of the Mesozoic Era (Upper Cretaceous, some 70 million years ago) the length of the day increased further to 23·67 hours (370·3 days in a year, 29·9 in a month) until the present values of 23·93 hours in a day, 365·26 days in a year and 29·53 days in a month were established by the beginning of the Quaternary Era.

The time which has elapsed since the beginning of the Palaeozoic Era represents a little more than 10% of the age of the Earth–Moon system. A backward linear extrapolation of the effects of tidal torque, at its level of the last 500 million years, to intervals several times that long ago in the past is probably invalid, for if we do so (H Gerstenkorn, G J F MacDonald and others), we find that the Moon should have been in the immediate proximity of the Earth 2·8 billion years ago. Such a close approach of both bodies would have left indelible marks on the surface of the Moon, but we know today that no noteworthy event occurred on the Moon at that time.

But while tidal friction in the Earth–Moon system may not have operated at a uniform rate in the past, its future is probably better predictable. At the present time the Moon is receding from the Earth by about 3·2 centimetres per annum; but even so, the relatively large store of angular momentum still available in the axial rotation of the Earth continues to represent a fair supply of momentum to predatory lunar tides. Their slow but relentless action will continue to lengthen our day and our month by a gradual transfer of the orbital momentum of the Earth to the orbital momentum of the system. This process can only come to a standstill when, finally, a synchronism between rotation and revolution will be attained not only for the Moon, but also for the Earth.

By the end of this final act of tidal evolution, the relative orbit of the Moon around the Earth will have become circular (its mean eccentricity at present is still 0·054) and approximately 1·59 times larger than it is at present. The tides raised by both bodies on each other will then become equilibrium tides which produce no torque, and in the absence of any outside dissipative forces the Earth–Moon orbit should become immune to any further change. The duration of the day on both the Earth and the Moon will become equal to their orbital period of approximately 55 days (or two months) of our present time, and the axes of rotation of both components will be perpendicular to their orbital plane—a development which will end the alternation of the seasons on Earth. The Moon will appear to us about one-third smaller in the sky than it is today and will continue to show us exactly the same, immutable face—as the Earth will to the Moon.

Not perhaps a particularly exciting end to the long romance which must have begun with a much closer gravitational embrace 4·6 billion years ago and one which (some would maintain) created much heat in both partners at that time. But such seems to be the end of most long-lasting liaisons, whether in the sky or on the Earth. The saving grace which may absolve our planet from its sheer monotony could only present itself if solar gravitational perturbations were eventually to bring about a formal dissolution of the partnership consistent with the laws of celestial mechanics. The only possible loophole is provided by the finite eccentricity of the terrestrial orbit: this allows, though not ensures, a leakage of energy and momentum and their gradual transfer from the Earth–Moon to the Earth–Sun system; while the latter is indissoluble, the former may not be. But it is certain that such an event, even if admissible under celestial divorce laws, would take a very long time—much longer than that vouchsafed to life on Earth in the future. Thus no-one, I am glad to say, will be around to see our Earth bid farewell to its old satellite which has illuminated its nights so faithfully and for so long.

4 Mercury and Pluto: The Sentinels of the Solar System

Next to the Moon in mass, Mercury and Pluto occupy positions of special significance in the solar system as its innermost and outermost planets. Mercury is situated in the scorching proximity of the Sun whose radiation warms its surface to almost red heat, while Pluto is situated at the outer limit of the system where the Sun appears only as a star in the daytime sky, its light barely interfering with the monotony of the cold and dark of interstellar space. But in spite of the difference between the extremes of location and climate, these two planets appear to have much in common from a dynamical, as well as a physical point of view—so much, in fact, that Pluto can be regarded as a 'stray' terrestrial planet separated from its partners merely by the belt of the giants. For this reason, we shall include its description here despite its location on the outskirts of the solar system. Pluto is the outer sentinel of the system, just as Mercury is the inner one.

Because of their extreme locations, both planets are difficult to observe from the Earth and offer their own peculiar difficulties to exploration. The existence of Pluto was discovered in 1930 by the young American astronomer Clyde Tombaugh at Lowell Observatory in Arizona—an institution dedicated by its founder, Percival Lowell, partly to this task. Whether this discovery was made accidentally, or as a result of Lowell's prediction of the planet's existence from gravitational perturbations of its nearest inner neighbour Neptune, constitutes a moot point on which expert opinions continue to differ, although these differences fortunately change nothing on the outcome. But with Mercury it was a different story. Although this planet is much nearer to us than Pluto and continues to bask in the intense glare of the Sun, its very proximity to the Sun never allows it to elongate from the Sun in our sky by more than 28°. It can, therefore, never be seen from the Earth except in the twilight hours shortly after sunset or before sunrise, a fact which makes its observation difficult. There is a story (probably apocryphal) that Copernicus lamented on his deathbed to have never seen it.

Both Mercury and Pluto revolve around the Sun in orbits more eccentric ($e = 0 \cdot 206$ for Mercury and $0 \cdot 249$ for Pluto) and more inclined to the ecliptic (by $7^{\circ}\!\cdot\!0$ for Mercury and $17^{\circ}\!\cdot\!1$ for Pluto) than those of any other planet in the solar system. The actual distance of Mercury from the Sun oscillates between 46 and 69 million km, and light traverses its mean distance of 57 million km in 3 minutes and 13 seconds. The mean distance of Pluto from the Sun amounts to 5910 million km, and light from the Sun needs on average no less than 6 hours and 28 minutes to reach this outermost planet.

The orbital periods of Mercury and Pluto differ accordingly. Whereas the *sidereal year* P_* for Mercury (i.e. the time taken by the planet to return to the same place in the sky) is only 87·696 mean solar days of terrestrial time, Pluto needs 349·17 years to accomplish one such orbit around the Sun. On the other hand, the *synodic year* P_s for Mercury (i.e. the time taken by the planet to exhibit the same phase to an observer on the Earth which is revolving around the Sun in the same direction with a period of $P_E = 365\cdot256$ days) is given by the equation

$$\frac{1}{P_s} = \frac{1}{P_*} - \frac{1}{P_E}, \tag{4.1}$$

and is equal to 115·404 days of our time. While the actual distance of Mercury from the Earth fluctuates during this time between 78 and 218 million km as the phase of the planet grows from 'new' to 'full', the brightness of Mercury varies between −1·2 apparent visual magnitude (equal almost to that of the star Sirius) and +1·1 apparent visual magnitude (comparable with that of Aldebaran), the maximum brightness occurring at a crescent phase between the 'new' and 'first' quarter. Although both Sirius and Aldebaran are conspicuous stars of our night-time sky, they are not conspicuous in the twilight hours or when they are just above the horizon; and, under these circumstances, nor is Mercury. Pluto is visible during the night at the time of its opposition and its apparent brightness of 14·6 visual magnitude (15·3 in photographic light) remains almost constant, so that quite a large telescope is required to see it.

Other important features which Mercury and Pluto have in common are their small size and mass. The apparent diameter of Mercury never exceeds 13 seconds of arc, and is difficult to measure accurately through a telescope because Mercury is so close to the Sun. The measurements of the apparent size of Mercury during its occasional transits across the Sun (occurring a few times per century) are again subject to their own peculiar errors. As long as telescopes constituted the only means of planetary exploration, the actual dimensions of Mercury remained more uncertain than those of any other planet (excluding Pluto), and estimates of its diameter ranged between 4800 and 4900 km.

Since 1962, however, Mercury has been contacted by terrestrial radars. By an accurate timing of radar echoes from the disc of the planet (i.e. a determination of the time-lag between the outgoing and returning signals) radio-engineers have found it possible to determine the distance separating us from Mercury with the precision far beyond that of any previous tele-

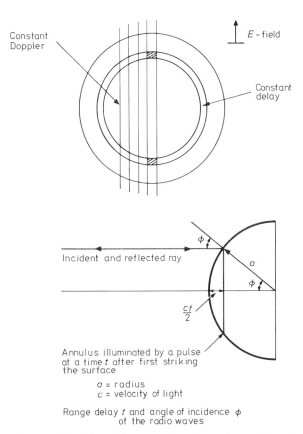

Figure 8. Principles of the range-Doppler radar tracking of planetary surfaces (schematic).

scopic triangulation. Moreover, an echo reaching us from a sphere will not be instantaneous but will possess a finitie spread, depending on the time-lag in its arrival from the element of the surface having the Earth at its zenith, and from the limb at which the Earth would be seen on the horizon (see figure 8). These two paths differ in length by the radius of the reflecting planet. The width of the echo then specifies the 'radar radius' or absolute dimensions of the spherical reflector.

An analysis of the radar echoes from Mercury has shown that the diameter of this planet is equal to 4879 ± 1 km, or 0·3829 times the mean diameter of the Earth. Subsequent direct measurements of this diameter by the Mariner 10 spacecraft in 1974–1975 (see table 6) have confirmed it to within the given uncertainty. As for Pluto, that planet has proved to be hopelessly beyond the range of any terrestrial radar signals, and the angular diameter of 0″.13 of its apparent disc as measured by telescope is still subject to considerable uncertainty. It corresponds, however, to an absolute value close to 6000 km—somewhat larger than Mercury's—and is unlikely to be improved much in the foreseeable future.

Like that of any other planetary body, the mass of Mercury or Pluto can be determined only from the effects of its attraction on the motion of some other body in its proximity, the 'perturbations' so caused being the larger, the larger the mass of the disturbing body. In the days of telescopic astronomy this used to be Venus (and, to a lesser extent, the Earth) for Mercury, and Neptune (or Uranus) for Pluto. The masses of both Mercury and Pluto are much smaller than those of their neighbouring planets, so that perturbations caused by their attraction will be small and difficult to detect. This is especially the case for Pluto, since its nearest planetary neighbour, Neptune, is much more massive and moves but slowly in the sky because of its great distance from the Sun.

The mass of Mercury required to account for the observed perturbations inflicted by it on the orbits of Venus or the Earth turned out to be about 1 part in 6 000 000 of that of the Sun (or 0·053 of that of the Earth) with an uncertainty of the order of 1–2%. A close approach to Mercury by the Mariner 10 spacecraft in 1974 provided a much more sensitive indication of the mass of the planet than its relatively large planetary neighbours. The outcome of the analysis of Mercury's attraction on the motion of Mariner 10 (tracked accurately by the range-Doppler signals transmitted by the spacecraft) indicated the mass of the planet to be equal to $0·055\,275 \pm 0·000\,003$ of that of the Earth, or $3·303 \times 10^{26}$ g. If we combine this mass with the diameter of 4879 km, the mean density of the Mercurian globe should be equal to $5·43 \pm 0·01$ g cm^{-3}, a value unexpectedly high for a planet of so small a mass and almost as large as that of the Earth. In addition, the same appears to be true of Pluto. Its mass of $0·11 \pm 0·02$ of that of the Earth (inferred from perturbations caused by it on the motion of Neptune) is the least well known of all the planets; and when combined with the estimated diameter of 6000 km, it leads to a density of 5–6 g cm^{-3}. The mean density of the Plutonian globe is, therefore, unlikely to be less than that of Mercury, and may even be higher.

Radar observations of Mercury can disclose not only its distance or size, but also the rate of its rotation and the orientation of its axis from the Doppler shifts of the returning radar echoes. If a spherical target bouncing back such echoes rotates, one limb of the sphere will be approaching us while the opposite limb will be receding (see again figure 8). This motion will, in turn, widen the echo of an (initially) sharp signal into a Doppler profile, the width of which will correspond to a difference in velocity of approach and recession at opposite limbs of the respective configuration. If the absolute size of the latter is known, the echo width can specify the corresponding period of axial rotation. It is true that this is possible from a single observation of this kind, made at a particular time, only if the orientation of the axis of rotation is known independently. In general, this will not be the case. However, at different positions of the planet in the sky its axis of rotation can be seen from different directions. The fluctuations in Doppler shifts caused by this phenomenon permit us to determine simultaneously the velocity of axial rotation and the orientation of its axis in space. As the corresponding Doppler shifts can be measured with relatively high precision, in many cases the radar studies of planetary rotation can lead to results that are far more accurate than those based on visual or photographic observations of surface markings.

The radar observations of Mercury—the surface of which does not exhibit any landmarks that could be reliably followed by a telescopic observer—have disclosed that this planet revolves about an axis virtually perpendicular to its orbital plane in a sidereal period of 58·65 days of our own time. The sense of its rotation is *direct*—that is, the planet rotates in the same sense as it revolves around the Sun. When we recall that Mercury's orbital period around the Sun is equal to 87·969 days, it transpires that this planet completes *three* rotations about its own axis while revolving *twice* around the Sun. That the periods of 58·65 and 87·97 days happen to be in the ratio of 2:3 within the limits of observational error is certainly no accident, but an indication of the existence of a spin–orbit coupling of tidal origin with the Sun as the tide-generating body. But why these periods became stabilized in the particular ratio 2:3 has not yet been satisfactorily explained.

For what period of time does the Sun remain continuously visible over any particular spot on Mercury's surface? This period D, known as the mean solar day, can be found from the equation

$$\frac{1}{D} = \frac{1}{58^{\text{d}}65} - \frac{1}{87^{\text{d}}97} \qquad (4.2)$$

to be equal to 176 days of terrestrial time, a time equal to three sidereal days of $58^{\text{d}}65$ on Mercury and to twice its sidereal year of $87^{\text{d}}97$. Thus, contrary to previous opinions, Mercury does not always show the Sun the same face. For an observer situated anywhere on Mercury, the Sun would rise in the east and set in the west every 176 terrestrial days (or 4224 hours) and would appear as a huge fiery disc of apparent diameter two to three times larger than we see it on Earth. At the time of each perihelion passage (i.e. when Mercury is closest to the Sun) the planet would pause in the sky and reverse its direction of motion for about eight days before resuming its westward advance, because its mean angular velocity of revolution exceeds that of axial rotation. This should cause excessive heating at the sub-solar points at that time, giving rise to *two* super-tropical regions situated on the planet's equator opposite to each other. Therefore, the *seasonal effects* on Mercury should be essentially longitudinal; on a planet whose axis of rotation is virtually normal to the plane of its orbit, those depending on latitude should be negligible.

The long duration of the solar day on Mercury alone would lead us to expect that the extremes of temperature on its surface between day and night are likely to be very large, an expectation amply borne out by the measurements of the intensity of infrared radiation from Mercury's surface, measured first (since 1962) from the Earth and later (since 1974) aboard the Mariner 10 spacecraft. In this way the temperatures at the sub-polar points of the Mercurian surface were found to be as high as 700 K, while at night they dropped to approximately 100 K. Thus, in spite of Mercury's proximity to the Sun, at night its surface becomes almost as cold as that of the Moon. (Nights on Mercury are 6·5 times longer than those on the Moon.)

The maintenance of surface-temperature differences of the order of 600 K implies a thermal inertia of surface material (consistent with a relative weakness of radar echoes returned by it) comparable with that on the Moon. From this alone we could surmise that the physical structure and chemical composition of the surface of Mercury should be similar to that of the Moon. In particular, this surface should be covered with the same kind of broken layer and by the same kinds of impact craters as found on the Moon.

These expectations were amply borne out when the first close-up pictures of the Mercurian surface were relayed to the Earth by the Mariner 10 spacecraft in March 1974 (plates 28 and 29). The sunlit part of the surface observed by Mariner 10 in the course of its first fly-by on 29 March is seen to be dominated by craters and basins very much like those seen on the Moon with young formations surrounded by streaks or aprons of bright ejecta. These are typical *impact* features very much like those on the Moon, though the similarity is, in fact, not complete. For instance, the heavily cratered regions on Mercury exhibit relatively smooth areas between craters, whereas the continental areas on the Moon show that craters with diameters greater than 20–30 km overlap. This means that the surface of Mercury has not been saturated with crater-forming bombardment to the same extent. This can be explained by the fact that the surface acceleration on Mercury ($3 \cdot 70$ m s^{-2}) is about twice as large as that on the Moon. As a result, the ejecta from the primary impacts have ballistic trajectories that do not reach as far and thus produce 'secondary' craters around the primary formation in groups more densely clustered than is the case on the Moon.

Another significant difference between the surfaces of Mercury and the Moon is the ubiquitous presence on Mercury of shallow cliffs which extend in places for hundreds of kilometres. The structure of these scarps suggests that they were formed by crustal shrinkage on a global scale. On the Moon, or Mars, such features appear conspicuous by their absence; instead, these bodies (of lower mean density) exhibit some effects of crustal stretching.

The 'outgoing' hemisphere of Mercury as photographed by Mariner 10 as it 'looked back' after its close approach (see plate 28) possessed surface features somewhat different from those on the 'incoming' hemisphere. Its surface is, in general, less heavily cratered, yet it exhibits the largest basins found anywhere on Mercury so far. One such formation,

Mercury and Pluto: The Sentinels of the Solar System 105

Plate 28. A mosaic view of the planet Mercury as recorded by Mariner 10 'looking back' after its historic fly-by on 29 March 1974 at a distance of approximately 210 000 kilometres from the planet. The Mercurian landscape recorded on this mosaic differs only in detail, but not in kind, from that to be found on the Moon: impact craters abound everywhere, some being surrounded (as on the Moon) by bright rays. The large circular basin ('Caloris Basin' only half-visible on the terminator near the centre of the disc), some 1300 kilometres in diameter, resembles the lunar Mare Orientale (see plate 17). North is to the top of the picture. *Photograph by courtesy of NASA.*

Plate 29. A more detailed view of the Mercurian landscape as recorded by Mariner 10 on 29 March 1974 during its first close approach to the planet. In the centre of the field can be seen a crater called 'Hun Kal' (signifying the number 20 in the language of the ancient Mayans) located approximately on Mercury's equator at about 20°W longitude. *Photograph by courtesy of NASA.*

1400 km long and referred to as the Caloris Basin (because of its equatorial location in one of the super-tropical regions), represents the scar of a gigantic impact comparable to that which gave rise to Mare Imbrium or Mare Orientale on the Moon.

A detailed study of the stony sculpture existing on the Mercurian surface bears out one general fact of over-riding importance: the far-reaching similarity between Mercury and the Moon suggests that the processes which shaped them were of the same kind. As the internal structure of these two bodies is quite different (we shall say more on this below), what else but external influences could have impressed so similar a stony structure on their surfaces? The similarity of impact cratering on Mercury or the Moon (and, as we shall see, on Mars) is surprising if we consider the very different locations of these bodies in the inner parts of the solar system and, in particular, their different distances from the asteroidal belt which is thought to be the principal source of the impacting bodies. Nevertheless, this seems to be the case.

The similarity between the surface structures of Mercury and the Moon goes even further. More detailed investigations show that the surface history of Mercury bears evidence of the same time-sequence of events as that which shaped the Moon. In particular, most events which left visible scars on the surface of Mercury seem again to have occurred within less than the first third of the planet's age; since that time conditions seem to have remained quiet, as they have been on the Moon. Like the Moon, Mercury continues to present a target to a diminishing number of impacts, and the craters created more recently can be distinguished from older formations by their bright ray systems. As on the Moon, they do not occur in large numbers, an independent testimony by Mercury to the fact that the rate of meteoritic bombardment in the solar system is gradually coming to a standstill.

If the surfaces of Mercury and the Moon exhibit such remarkable similarity, the same is *not* true of their internal structure. The fact that Mercury as a planet with a mass only 4·5 times larger than that of the Moon possesses a mean density of 5·43 g cm^{-3} in contrast with the lunar value of 3·34 g cm^{-3}, discloses that the material in its interior must be very different from that constituting the Moon, for the difference in mean densities is far beyond the range which could be accounted for by different self-compression. In addition, the relatively high mean density of the Mercurian globe—the second highest encountered among the terrestrial planets —requires the presence in its interior of an iron core, which, if similar in composition to that of the Earth, must extend (on account of the smaller degree of compression of its material) to approximately 1800 km—that is, to about 74% of Mercury's radius. The silicate mantle 'floating' on top of it cannot be much more than 600 km deep.

An indirect, though eloquent, proof of the existence of Mercury's iron core has been provided by Mariner 10's discovery that Mercury exhibits a dipole *magnetic field*, the axis of which is inclined by 12° to the spin axis of the planet. The equatorial strength of this field, as registered by a magnetometer aboard Mariner 10 during its repeated fly-by's of the planet, was found to correspond to 330–350 gammas (1 gamma = 10^{-5} gauss) or to about 1% of the strength of the terrestrial magnetic field. Mercury's field is, however, much stronger than those of the Moon, Venus, or Mars (which for the first two are beyond the limits of detectability) and an internal global mechanism seems to be required for its generation. The magnetic field of the Earth—by far the strongest generated by any terrestrial planet—is supposed to have originated from the electric currents produced by the axial rotation of a fluid metallic core ('dynamo effect'). The planet Mercury no doubt possesses a similar core of even larger fractional dimensions than that of the Earth, but its rotation is 59 times slower. Is this sufficient to produce a field of 0·003–0·005 gauss by dynamo action?

Another explanation for such a field could be based on the surmise that it represents a 'fossil' field acquired in the early days of the solar system by an as yet unknown process. But it seems unlikely that, in more than the four billion years which have elapsed since that time, the temperature of Mercury's interior never rose above the Curie point (the value where any substance loses its magnetism). A third possibility is the acquisition of a magnetic field by induction from the solar wind, but this again would scarcely endow the induced field with an inclination to the axis of rotation which this field possesses. Perhaps the magnetic field of Mercury arises from causes as yet unknown, or perhaps we shall have to gain a deeper understanding of the mechanism by which our Earth generates its own field before this can be applied to Mercury. Whatever the case may be, it is fortunate that a magnetic field has been discovered in at least one other terrestrial planet with which we can compare the field of the Earth.

Another important contribution by Mariner 10 was the detection of a tenuous *atmosphere* surrounding Mercury and consisting largely of helium, with weak indications of neon, argon, and (possibly) xenon over the dark side. Its primary source is probably external, and, as on the Moon, most of it represents the 'solar wind' entrapped by the planet's magnetic field and recombined into neutral gas by collisions with the planet's surface. Emanations from the surface due to radioactive disintegrations may constitute its second significant source. The mean density of such particles detected by Mariner 10 is, however, so small as to create a pressure on the surface of the order of only 10^{-9} millibars (corresponding to one-trillionth of an atmosphere), so that, as around the Moon, the individual particles of so tenuous a gas only seldom collide and for most of the time move along 'ballistic trajectories' in the planet's gravitational field. Such an assembly of particles scarcely deserves to be called an atmosphere—'exosphere' would be a better description. It is far too tenuous to protect the surface of Mercury from cosmic impacts of any size (including those of interplanetary dust), or to produce any appreciable erosion of the surface over astronomically long periods of time.

In bidding farewell to the planetary messenger of the gods at this stage of its exploration, let us close the present chapter with a few additional words on Pluto (see plate 30). This outermost planet of the solar system is likewise found to rotate about an axis (of unknown orientation) in a much shorter period than Mercury and in a time amounting to 6 days, 9 hours and 17 minutes, as evidenced by small but measurable changes in its brightness fluctuations during that period. The fluctuations are no doubt produced by successive meridional passages of regions reflecting sunlight to different extents. Pluto is too far away—and, therefore, its apparent disc too small—for us to see such regions as bright or dark spots, but photometers have disclosed their alternation even though the size of these spots is below the limit of resolution of our telescopes.

Nothing whatever is known so far about the structure of the

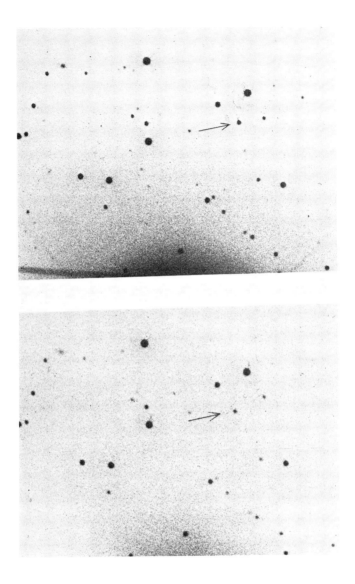

Plate 30. Two photographs of the planet Pluto (marked by arrows) as a star of fifteenth magnitude taken on 22 and 23 March 1930 (the year of its discovery). The planet's apparent motion with respect to the neighbouring stars is clearly noticeable in the course of one day. The bright glare near the bottom of each frame is that of the naked-eye star δ Geminorum. Photographs taken with the 60 inch reflector of Mount Wilson Observatory.

surface or about the interior of Pluto. At its mean distance of 39·5 AU from the Sun, in its sky the Sun would not appear any larger than the planet Jupiter appears to us on the Earth, and a telescope (albeit a small one) would be required to disclose the finite angular size of the Sun's disc. Its light is so diluted by distance that even in daytime absorption would not warm the surface of Pluto to more than 40–50 K. How close the actual temperatures are to this limit remains to be verified by actual observations.

Colour plate 1 (a)

Colour plate 1 (*b*)

Colour plate 2

Colour plate 3

Colour plate 4

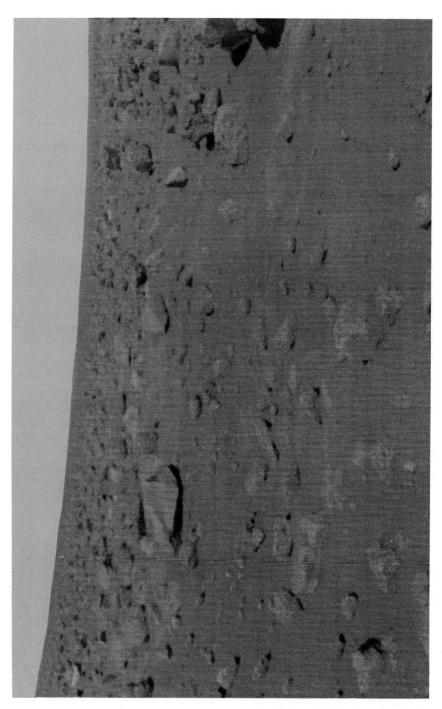

Colour plate 5

Colour plate 6 (a)

Colour plate 6 (*b*)

Colour plate 7 (a)

Colour plate 7(b)

Colour plate 8

Colour plate 9

Colour Plates

Colour plate 1(a). Earthrise over the lunar landscape, as photographed from the Apollo 11 command module in orbit around the Moon in July 1969. *Photograph by courtesy of NASA.*

Colour plate 1(b). The command–service module of the Apollo 17 mission in orbit around the Moon with its scientific instruments module bay exposed to view. *Photograph by courtesy of NASA.*

Colour plate 2. Apollo 15 lunar roving vehicle (1971 model). *Photograph by courtesy of NASA.*

Colour plate 3. Apollo 17 scientist–astronaut Dr Harrison H Schmitt (now US Senator for New Mexico) investigating the lunar landscape in the Taurus–Littrow region in December 1972. *Photograph by courtesy of NASA.*

Colour plate 4. Apollo 15 lunar excursion module *Falcon* at the foothills of the lunar Apennines in July 1971. The Mission Commander, astronaut D R Scott, is on the left and the lunar roving vehicle on the right of the LEM. Mount Hadley, the highest peak of the Apennines and one of the highest mountains on the Moon (rising to almost 5000 metres above the surrounding landscape), looms in the background. *Photograph by courtesy of NASA.*

Colour plate 5. The Martian landscape in the afternoon and in a north-easterly direction as televised by the Viking 2 spacecraft on 5 September 1976. The salmon-pink colour of the landscape is due to an overlay of dust consisting mainly of hydrated ferric oxide (limonite). A fine dust of the same substance stirred by convection is responsible for the similar colour of the daytime sky. The horizon appears tilted because the spacecraft landed with its cameras inclined approximately 8° to the west. *Photograph by courtesy of NASA and JPL.*

Colour plate 6(a). The Martian landscape in the region of Chryses Planitia as televised by the Viking 1 spacecraft on 15 August 1976. The horizon is at a distance of approximately 3 kilometres from the spacecraft, parts of which can be seen in the foreground. On the right is the retractable scoop in its 'parked' position; the magnet-cleaning brush mounted at the end of a short boom can be seen on the left. The box on the lower right housed the meteorological instruments prior to their deployment after landing. *Photograph by courtesy of NASA and JPL.*

Colour plate 6(b). A bird's-eye view of the Martian surface from an altitude of 31 000 kilometres and covering an area of approximately 1800 × 2000 kilometres, as televised by the Viking 1 orbiter on 22 July 1976 from a vantage point near the apoareum of its orbit. The principal feature seen on the ground is part of the great Valley of the Mariners which runs parallel to the Martian equator and is almost 5000 kilometres in length. *Photograph by courtesy of NASA and JPL.*

Colour plate 7(a). A view of the Martian landscape in the afternoon and in the

Colour Plates

neighbourhood of the landing site of the Viking 2 spacecraft, as televised on 7 September 1976. Parts of the spacecraft can be seen in the foreground. The salmon-pink colour of the landscape and of the daytime sky is caused by an overlay of the dust of hydrated ferric oxide. Note in the picture, however, that the steep sides of the boulders (to which dust can no longer adhere) are much darker, indicating a basaltic composition. *Photograph by courtesy of NASA and JPL.*

Colour plate 7(*b*). Sunset over the Martian landscape, as televised by the Viking 2 spacecraft in September 1976, showing the streamers of the solar corona (see also plate 5) protruding above the horizon after the Sun's disc has set. Note that the colour of the Martian sky (salmon-pink during the daytime; colour plates 5 and 7(*a*)), reverts to a deep blue (due to molecular scattering) after the diurnal atmospheric convection currents cease to stir and raise surface dust. *Photograph by courtesy of NASA and JPL.*

Colour plate 8. A view of the Earth as photographed at a distance of 172 000 kilometres by the Apollo 11 mission on its outward journey to the Moon. The African continent and the Arabian peninsula with adjacent parts of the Middle East can be seen quite clearly. *Photograph by courtesy of NASA.*

Colour plate 9. A rare view of part of South-East Asia taken from the Apollo 7 spacecraft in orbit around the Earth and showing the vast plains of Tibet (right) separated from the jungles of Nepal (left) by the magnificent mountain chain of the Himalayas (extending in the picture from the Kangchenjunga and Everest groups to the high peaks of Kashmir). The photograph was taken while the camera was in orbital motion at a speed of almost 8 km s^{-1}. However, some 200 kilometres below, a much more dramatic event was unfolding at a speed lower by 14 orders of magnitude: namely, the collision between the main Asiatic land mass and the indian subcontinent, proceeding with its momentum unchecked at a speed of several millimetres per year. The consequences of this collision have already raised the Himalayas to their unprecedented heights and continue to push them even higher. *Photograph by courtesy of NASA.*

5 Mars: The Portrait of a Midi-Planet

Mars is our second-nearest planetary neighbour and also the second planet on which man-made spacecraft have recently effected soft landings. Apart from the Earth and the Moon, Mars has been the first planet to be provided with several artificial satellites sent out from the Earth on reconnaissance missions. Until quite recently Mars was the last candidate in the solar system as a possible indigenous home for life as distinct from that which has developed on the Earth, and of which we are the crowning achievement. Although the hope for life on Mars has been shattered by the outcome of the Viking missions of 1976, and the interest in Mars has thus diminished for the biologists, other aspects of this outcome have made Mars no less interesting to the astronomers, physicists, inorganic chemists and geologists who seek a deeper understanding of the evolution of other planetary surfaces besides that of the Earth.

Facts and Figures

Let us first introduce Mars as an astronomical body and discuss its location within the solar system as well as its kinematic characteristics. It revolves around the Sun in an orbit whose size places this planet well outside that of the Earth. Mars is therefore an *outer planet* which can be observed throughout the night; its apparent brightness around the time of its closest approach to us can make it one of the brightest objects in our sky. In more specific terms, its mean distance from the Sun is equal to 1·524 AU and it revolves in an orbit of marked eccentricity ($e = 0·093\ 26$) which is, however, not as large as that of Mercury and never brings Mars closer to the Earth than 55·5 million km. This can happen only when Mars is in 'opposition' to the Earth (i.e. when the Earth is between the Sun and Mars) at a time when Mars is at perihelion (i.e. at its nearest to the Sun) while the Earth is at aphelion (i.e. at its farthest from the Sun), a geometrical configuration which occurs every 15³/₄ years. If Mars is on the other side of the Sun, its distance from us can increase to 378 million km.

In accordance with the size of its orbit, Mars revolves around the

Sun in a sidereal period P_* longer than one year—in fact a period equal to 686 days and 22·3 hours of our mean solar time. The time interval between successive oppositions or conjunctions, which is the synodic period P_s, follows from the equation

$$\frac{1}{P_s} = \frac{1}{365\overset{d}{\cdot}26} - \frac{1}{686\overset{d}{\cdot}93}, \qquad (5.1)$$

which gives a value for P_s of just over 780 days of our time. Therefore, the Earth needs more than two years to catch up with Mars in space, and the aspects of Mars as seen in solar illumination will repeat themselves to us in that period.

Even when Mars is at inferior conjunction (i.e. at its closest to us), its distance from the Earth is still about 1·5 times as large as that of Venus at inferior conjunction. As a result, spacecraft sent out to Mars (see table 6) have to spend some seven months or more en route—as opposed to a three-day trip to the Moon or a three-month one to Venus—before they reach the proximity of their target. This is due to the fact that a spacecraft on its way to Mars spends most of its time travelling against the direction of solar attraction, while for travel to Venus (an inner planet) the opposite is true.

The relative sizes of the terrestrial and Martian orbits are such that the apparent disc of Mars, as seen from the Earth, never departs from 'full' phase by more than 47° in phase angle. Mars can therefore appear distinctly gibbous† to us through the telescope, but never as a crescent. Since the maximum apparent diameter of Mars at inferior conjunction never exceeds 25·1 seconds of arc, a telescope is necessary to see Mars in a form other than as a light point. At the minimum distance of 55·5 million km this apparent diameter corresponds to a configuration with a mean radius of 3400 km, a value which recent spacecraft data have refined to 3390 km with an uncertainty of less than ±1 km. In size, therefore, Mars is slightly larger than one-half of the Earth and slightly smaller than twice the size of the Moon. By size it occupies a position half-way between the sizes of these two bodies which represent examples of the largest and smallest terrestrial planets, respectively. Such a position entitles us to regard Mars as a 'midi-planet'.

As stressed already, the mass of any celestial body can be inferred only from the effects which its attraction exerts on any other nearby mass. Unlike Mercury or Pluto, for Mars we do not need to base such a determination on the perturbations caused by its mass to the motion of its neighbouring planets—the Earth or Jupiter; the latter especially is too far away and too massive to be affected by its small neighbour. Since 1877 we have known that Mars possesses two natural satellites, but both of these are

---

†A phase of an illuminated sphere between 'full' and the first or last quarter.

too small and too difficult to observe to serve as good indications of the mass of Mars. Fortunately, in recent years this mass has been deduced with much greater precision from the observed motions of fly-by spacecraft (see table 6) which can be tracked by radio methods using range-Doppler techniques. In particular, from the motion of the early Mariner spacecraft, the ratio of the mass of Mars ($m_\male$) to that of the Sun ($m_\odot$) was determined to be

$$\frac{m_\odot}{m_\male} = 3\ 098\ 600 \pm 600, \tag{5.2}$$

which shows that the Sun is more than three million times as massive as Mars. In terms of the mass of the Earth ($m_\oplus$),

$$\frac{m_\male}{m_\oplus} = 0\cdot107\ 45 \pm 0\cdot000\ 02, \tag{5.3}$$

which shows that the mass of Mars amounts to only 10·7% of that of the Earth or, in absolute units, to approximately $6\cdot421 \times 10^{26}$ g. If we divide this mass by the volume of a sphere of mean radius 3390 km, the mean density of the Martian globe proves to be 3·95 g cm^{-3}, a value appreciably lower than that of Mercury and only 18% higher than that of the Moon. The gravitational acceleration $g_\male$ on the Martian surface is equal to 3·73 m s^{-2} (i.e. 0·38 times that on the Earth) and corresponds to a velocity of escape of only 5·03 km s^{-1}.

Since the early days of telescopic astronomy, the apparent disc of Mars has disclosed to terrestrial observers unmistakable evidence of a diurnal *rotation* much faster than Mercury's and with an axis of rotation inclined by 25°.2 to a direction perpendicular to the Martian orbital plane (for the Earth this inclination is equal to 23°.45). Extensive observations carried out for more than three centuries have revealed that the sidereal day on Mars is equal to 24 hours, 37 minutes and 22·667 seconds of our time, and that the sense of rotation is direct. A sidereal day on Mars (i.e. the time interval between successive passages of the same star through the Martian meridian) is therefore only 41 minutes and 19 seconds longer than that on the Earth. The Martian solar day (i.e. the time interval between successive passages of the Sun through the meridian) is longer than the sidereal day by only 2 minutes and 12 seconds, in contrast with a difference of 3 minutes and 56 seconds between the solar and sidereal day on the Earth.

As on the Earth, the rotation of Mars has given rise to a centrifugal force which has flattened its globe to a small but measurable extent. The ratio of the polar to the equatorial semi-axis differs from unity by approximately half a per cent. This means that the polar semi-axis is about 17 km shorter than the equatorial one, a fact which causes the planet to precess in space with an estimated period of close to 173 000 years.

The Martian Environment: Its Climate and Atmosphere

If we were to land on Mars, what kind of landscape could we expect to find? Since the seventeenth century the most conspicuous semi-permanent markings discernible on the surface by telescope have been the *polar caps*. These are extensive regions surrounding the poles of the planet and have a whitish tinge which contrasts distinctly with the reddish colour of most of the other parts of the surface. Ever since the existence of these caps was first noted by Fontana in 1636, their extent was found to be dependent upon the seasons of the Martian year: they increased with the approach of winter and shrank with the advent of spring on the respective hemisphere. In winter, the cap surrounding the north pole of the planet descends to an areographic latitude of 60–65 degrees and does not disappear even at the height of the summer season. On the other hand, the southern cap is never as large and may vanish completely in the summer. The regularity in advance and retreat of the polar caps suggests that the substance of which they are made melts (or rather sublimes) with the advent of the Martian spring and solidifies again in the autumn.

The nature of the substance constituting these caps—the main candidates being solid water (ice) or carbon dioxide (dry ice)—was not resolved until the advent of spacecraft, and the answer almost proved to be an anticlimax. But before we unfold the solution, let us stress that the existence of polar caps and their seasonal variability proves that Mars, unlike Mercury or the Moon, does not expose a bare surface to interplanetary space: it is surrounded by a gaseous *atmosphere*, which offers the surface at least a limited protection from certain types of cosmic processes and enables others to leave distinct imprints.

That Mars is surrounded by an atmosphere of significant air mass was known a long time ago when the principal tools of exploration were still ground-based telescopes. Perhaps the simplest indication of the existence of an atmosphere was the fact that the planet's disc as seen (or photographed) through a telescope did not appear to be uniformly bright, but is darkened progressively towards the edge. In addition, the light of the planet was found to be partly polarized, the degree of polarization increasing between centre and limb. These phenomena indeed suggested the presence of an atmosphere of a non-negligible density, whose absorption and scattering dimmed the surface regions illuminated by the low Sun. More localized obscuration phenomena have also been observed to impair occasionally the visibility of surface details in certain regions, suggesting an interposition of different kinds of *clouds* ('white' or 'brown') which form and dissolve in the course of time. Such a process would again be impossible in the absence of an atmosphere.

How dense is the Martian atmosphere, and what does it consist of? That its principal constituent is carbon dioxide (CO_2) was discovered by astronomers (G P Kuiper 1947) from the characteristic absorption bands which this gas impresses upon light passing through it in the red and near-infrared regions of the spectrum. The discovery of other constituents of the atmosphere, as well as of the pressure which it exerts at different altitudes above the surface, had to await the advent of the space age. Thanks mainly to the contributions made in 1976 by the Viking spacecraft, we know today that carbon dioxide constitutes close to 95% of the Martian atmosphere by mass, nitrogen 2·7%, and argon (mainly radiogenic and originating from the beta-decay of ^{40}K), 1·6%. These three gases constitute more than 99% of the Martian air mass by weight. Oxygen (both atomic and molecular) amounts to 0·15%, and water vapour present in the Martian atmosphere corresponds to only 0·1 millimetres of precipitable liquid.

The total pressure of 7·6 ± 0·2 millibars exerted by all these constituents on the surface is subject to seasonal variations and diminishes with height more slowly than does our own atmosphere (because of the lower gravity on Mars), so that at altitudes above 90 km the Martian atmosphere is more dense than our own. We may also add that, even at the distance of Mars, the intensity of sunlight is sufficient to ionize the gases of the Martian upper atmosphere and thus sufficient to endow this planet with an *ionosphere*. According to the measurements made by successive Mariner spacecraft just before and after their occultation behind the limb of the planet, the maximum density of free electrons liberated by ionization is encountered at about 120 km above the Martian surface (in contrast to 300 km on the Earth). Because of the greater dilution of incident sunlight at the distance of Mars, this density amounts to only about 10^5 electrons per cm^3, a value about one-tenth of their corresponding number in our ionosphere.

But now let us descend from the ionospheric altitudes to the surface and return to the problem of the Martian polar caps (see plate 31). Both parent gases of 'wet' and 'dry' ice were found to be present in the Martian air, the latter in great abundance and the former in small but measurable amounts. In what proportion do their solid phases participate in the caps, and to what thickness do they add up? The reasoning by which these points can be established from observations is simple and depends essentially on the observed rate of melting (i.e. the retreat of the cap with advancing season), the amount of heat supplied by the Sun, and the melting point of the relevant substances at the prevalent pressure.

The rate of intake of solar heat per unit time at the distance of Mars is well known and is equal to approximately $(1·5)^2$ times less than that received by the Earth. The temperature to which this heat can raise the surface can be computed if the reflectivity and thermal inertia of the surface material are known. It can also be verified from measurements of thermal radiation emitted by the surface in the infrared and microwave regions of the spectrum. That these temperatures should be generally lower on Mars than on the Earth is to be expected because of the greater distance of that planet from the Sun, but only quantitative measurements can indicate their actual values.

116 *The Realm of the Terrestrial Planets*

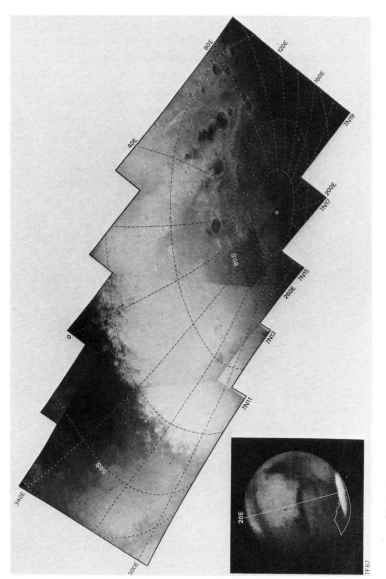

Plate 31. A photographic reconstruction of the southern polar cap on Mars from the data secured by the Mariner 7 spacecraft in August 1969. *Photograph by courtesy of NASA and JPL.*

The measurements of the intensity of infrared radiation from the planet's surface in the wavelength region between 8 and 12 μm (not visible to the eye, but one to which our atmosphere happens to be transparent) by W M Sinton and J Strong (1960) or E Pettit (1965) indicated that temperatures close to 0 °C were reached at the sub-solar point in the Martian tropics and fall to -20 to -30 °C at the time of the sunset. At the rims of the polar caps temperatures as low as -60 °C were registered, values sufficiently low to freeze water but not low enough to freeze out carbon dioxide. Night-time temperatures on Mars, or on any outer planet, cannot of course be measured directly from the Earth, for only a small part of the night hemisphere becomes visible to us by virtue of the phase effect. However, at least an indication of these can be obtained from measurements of the thermal emission of the Martian surface in the range of the radio-frequencies (i.e. at wavelengths between 0·1 and 10 cm), to which our atmosphere is again transparent, and which originate at depths equal to several times the wavelength. At these wavelengths, the emission by the Martian disc appeared to be essentially constant and indicative of a mean temperature between -70 and -80 °C. As the night-time temperatures on the exposed surface are bound to be even lower, they were estimated to be in the neighbourhood of -100 °C, a value almost sufficient to freeze out carbon dioxide from the Martian air.

The Mariners and Vikings

The historic missions of Mariners 6 and 7 to Mars in 1969 (followed by others listed in tables 6 and 7) and the soft landings of Vikings 1 and 2 on the surface in the summer of 1976 established the diurnal variations of temperature at their respective landing sites with a precision far exceeding that of the previous measurements made at a distance. On 20 July 1976, Viking 1 landed in the outskirts of the Martian tropics at a site of areographic latitude 23°4N, while two months later (on 2 September) Viking 2 landed at a site of 48°0N. The range of temperatures recorded by them came out to be lower than expected: a maximum of about -30 °C was reached, not at noon, but at about 3.30 pm Martian time, while the night-time temperatures fell to no lower than approximately -90 °C just before sunrise. However, a diurnal variation of 60 °C in 12 hours corresponds to an average temperature gradient of only 5 °C per hour (steeper at times), an amount much larger than on Mercury or the Moon because the Martian day is so short.

The Viking landers could only record the local temperatures prevalent at their respective landing sites. Temperatures over larger areas were established from infrared measurements made aboard the orbiting parts of these spacecraft. In particular, on the rim of the north polar cap the ground temperature proved to be only -120 °C with an uncertainty of the order of ± 10 °C. If we compare this value with the temperature of -125 °C

Table 7. Soft landings on Mars

Mission	Date of launch	Date of landing	Areographic coordinates of landing sites		Results
			Longitude	Latitude	
Mars 2	19 May 1971	27 November 1971	Unknown		No signals
Mars 3	28 May 1971	2 December 1971	$\lambda = 158°W$,	$\beta = 45°S$	Video signals ceased after 20 seconds
Mars 6	5 August 1973	12 March 1974	$\lambda = 25°W$,	$\beta = 24°S$	No signals
Viking 1	20 August 1975	20 July 1976	$\lambda = 48°.05W$,	$\beta = 22°.48N$	Excellent
Viking 2	10 September 1975	3 September 1976	$\lambda = 225°.74W$,	$\beta = 47°.97N$	Excellent

at which carbon dioxide solidifies (at a pressure of 7 millibars), the message of the measurements becomes unmistakable: the bulk of the material constituting the Martian polar caps is not ice—which would have remained solid at temperatures much higher than $-120\,°C$—but frozen carbon dioxide. Thus, in spite of their superficial similarity at a distance, in reality the Martian polar caps are vastly different from the massive fields of glaciers covering the polar regions of the Earth. The Martian polar caps consist largely of frozen layers of dry ice—no doubt mixed with some frozen water—and their bulk is formed by carbon dioxide. This fits in with the relative proportions of gaseous CO_2 and H_2O in the Martian atmosphere with its preponderance of CO_2 and scarcity of H_2O. Hence one of the age-old problems of the Martian environment appears to have been finally solved.

The Martian 'snowfields' in the polar regions (see plate 31) must be less than a few centimetres deep, for otherwise they could not form or disappear as rapidly as is observed at different seasons of the year. They are indeed a far cry from the massive ice caps of the terrestrial Arctic or Antarctic regions which have been stable over geologically long periods of time! In contrast, the Martian polar caps tend to 'grow from the air'. With the advent of the winter season on Mars, the polar temperatures become low enough for at least some gaseous CO_2 to freeze out of the atmosphere and cover the ground with a hoar-frost of dry ice which starts to vanish again (partly in the north, more completely in the south) with the coming of spring. That this occurs has been confirmed by barometric measurements made by the Viking landers: as the winter approached, the air pressure registered by them diminished because of a gradual conversion of atmospheric CO_2 into its solid phase.

How much cooler would it have to become on Mars for the bulk of CO_2 to solidify from the atmosphere and coat the entire planet with a layer of dry ice? The answer is—not much, and this would probably happen if Mars were to undergo an ice age similar to those that have occurred on the Earth in the relatively recent past (p 215). In the course of the terrestrial Quaternary ice age the global temperature of our planet dropped repeatedly by $10–15\,°C$ below its present level. If the same thing had happened on Mars at the same time, there is no doubt that 95% of its atmospheric air mass would have solidified on the surface, leaving only a tenuous envelope of argon and nitrogen to protect the Martian surface from the vicissitudes of interplanetary space. Whether a global glaciation by dry ice could have left any landmarks on the solid surface of Mars which we could decipher today is a tempting question, but at the moment we do not know the answer.

Moreover, so low is the atmospheric pressure on Mars that, with the advent of spring, neither solid CO_2 nor H_2O ever melts to pass through a liquid phase, but sublimes directly back into the air. Therefore, no liquid can flow on the surface of Mars, either under the present conditions or under any conditions similar to those which now obtain. And if this has always been the case, none of the features observed on the solid surface could have been

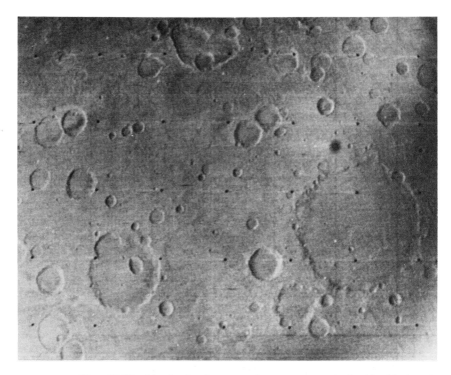

Plate 32. The Martian landscape at close range as televised by Mariner 6 at the time of its close approach to Mars on 30 July 1969. The view covers an area of approximately 700 × 900 kilometres in size which abounds with shallow crater formations of impact origin, the largest (to the right) being 260 kilometres across. *Photograph by courtesy of NASA and JPL.*

caused, or modified, by liquid flow. As we shall see later, this raises other problems which remain as yet unresolved.

While the Viking landers have transmitted extensive meteorological data characteristic of their landing sites, in the course of their descent to the surface they kept a continuous record of not only the pressure, but also of the temperature aloft. The data disclosed that, at an altitude of no more than 20 km above the surface, the daytime temperature had already fallen to about $-90\,°C$ and, at 50 km, to $-140\,°C$, none of which is below the CO_2 condensation boundary at the prevalent pressure. A minimum of $-150\,°C$ was recorded at an altitude of 70 km, but thereafter the temperature began to rise, attaining $+200\,°C$ in the ionosphere. Even this is much lower than the temperatures in excess of $1000\,°C$ recorded in the corresponding layers of our terrestrial atmosphere.

Many a reader may have experienced temperatures of -30 or $-40\,°C$, and temperatures as low as $-70\,°C$ have been recorded in Siberia, Alaska and the Antarctic. But what would it feel like to experience a temperature of +1000 °C in the Martian ionosphere? The answer is—still very cold, the reason being the low pressure and density of the ionosphere. A temperature of 1000 °C is merely a measure for the mean velocity with which atoms or molecules move in such a medium. Its individual particles are fast, but at the low pressure obtaining in the Martian ionosphere there are so few of them (per unit volume) that they would not affect the temperature sensors of our nervous systems. In other words, while the energy of individual particles may be high, this fact means nothing to our nervous systems unless the energy density is sufficiently high; this is no less true in the Martian ionosphere than in the ionosphere of the Earth.

The Viking landers managed to descend from the high altitudes of the Martian ionosphere to the surface in 11 minutes. Let us now describe the features of their landing sites as seen through the television cameras aboard these spacecraft. Colour plates 5, 6(a) and 7(a) record the views televised to the Earth through the intermediary relay of the Viking orbiters and illustrate a landscape very much rougher, on the centimetre to metre scale, than the corresponding lowlands on the Moon (cf colour plates 1(a), 3 and 4). The colour of the surface of the Martian landscape fully confirms that disclosed previously by telescopic observations, and the x-ray fluorescence spectrometer aboard the Viking landers established the chemical composition of the surface material beyond any doubt. In agreement with previous expectations, the material turned out to be mostly hydrated ferric oxide, a substance well known to mineralogists as limonite (Fe_3O_4). The colour (reflectivity) of this compound—responsible for the reddish tinge of desert areas on the Earth (see colour plates 8 or 9) or for the red colour of ordinary bricks—dominates the colour of the entire planet Mars as seen by a distant observer, and the ubiquity of this substance on the Martian surface evidences a high degree of oxidation of this surface.

The preponderance of limonite over the Martian landscape is, however, evident only in the small-grain component of its topmost cover and does not extend to stones and boulders whose sizes are on the centimetre or metre scale. A glance at the photograph in colour plate 7(a) illustrates that when the steep slopes of such boulders prevent finer dust from adhering to their faces, their general colour is very dark and indicative of a different type of composition. Although none of them was analysed chemically by the Vikings, a presumption is strong that their material is similar in composition to terrestrial or lunar basalts, and possibly of volcanic origin.

The horizontal panorama, as seen by the Vikings and shown in plates 33 and 34, also shows distinct features that are completely absent from the Moon: namely, dunes of sand or dust accumulated (or transported) by the action of *winds* in the Martian air. Indeed, anemometers aboard Viking 1 registered the almost continuous presence of winds blowing over the Martian landscape (mainly from the south west) with velocities between 3 and 30 km h^{-1}—on the whole, a moderate breeze. Occasional gusts of wind up to 50 km h^{-1} have been recorded. Evidence provided by dust storms

Plate 33(a). A panoramic view of the Martian landscape towards the east as televised by the Viking 1 soft-lander on 3 August 1976. The horizon is approximately 3 kilometres away and sand dunes stand out plastically in the morning illumination (the Sun is about 30° above the horizon). The large boulder to the left, playfully called 'Big Joe', is approximately 8 metres from the spacecraft and 3 metres across. The boom cutting the view in the middle belongs to the Viking meteorological sensor (cf figure 4). *Photograph by courtesy of NASA and JPL.*

Mars: The Portrait of a Midi-Planet 123

Plate 33(b). A high-resolution view of the Martian ground (and of one of the landing pads) transmitted only minutes after the soft-landing of Viking 1 on 20 July 1976. The largest rocks in the view are only about 10 centimetres across; the smallest details are less than a millimetre in size. *Photograph by courtesy of NASA and JPL.*

has also revealed that large parts of the planet's surface may be swept by much faster winds. These stir the surface dust into clouds which blanket large parts of the surface for weeks at a time and reduce the visibility of the surface details to zero. Such a great dust storm occurred on Mars and raged for several weeks at the time of the visits of Mariner 9 in December 1971 and in early January 1972. The winds of the storm raised enough dust to obliterate the visibility from space of virtually all the surface markings, and their velocity was estimated to be between 100 and 200 km h^{-1} at temperatures comparable to those found only in the Antarctic—not exactly the weather for outdoor walking! Thus, in spite of a relatively low density of Martian air above the surface, dust stirred up and transported by winds of this force can not only produce dunes as seen in plate 33(a), but can also etch the surface of boulders too heavy to be moved with innumerable pits centimetres to millimetres in size, as seen in plates 33(b) and 35 in the immediate proximity of the spacecraft. Whatever the final verdict may be on the erosion of the Martian surface by water in the past, the continuing operation of erosion by winds is undoubted; its effect can be seen in the eroded surface structure of Martian boulders (plate 35) as well as in the dunes accumulated on the horizon (plate 33(a)).

The winds sweeping over vast expanses of the Martian surface in a horizontal direction are not the only meteorological disturbances in the Martian environment: another contribution comes from the diurnal convection in the Martian air caused by the heating of the surface of the planet during daytime. We have mentioned earlier in this chapter that the temperature of the surface (about -85 °C just after sunrise) rises to -30 °C in the mid-afternoon. Such an increase will, in turn, heat the air immediately above the surface and cause it to rise. Air motions resulting from this convection are moderate, but nevertheless sufficient to stir dust on the surface and carry it aloft, just as similar convection currents in the terrestrial atmosphere carry aloft water vapour from the surface to form cumulus clouds in the air. On Mars the medium transported in this manner is dust formed predominantly by limonite. The scattering of sunlight on the dust is mainly responsible for the salmon-pink colour of the Martian sky during daytime (colour plates 5, 6(a) and 7(a)). If there were no dust in the Martian air, the colour of its sky would be deep blue to dark—as at high altitudes above the surface of the Earth—due to the Rayleigh scattering of sunlight on air molecules. This is indeed true on Mars at the time of sunrise or sunset (colour plate 7(b)) before air convection begins, or after it has ceased, to operate. However, less than an hour after sunrise there is enough dust in the air to endow the Martian sky with its salmon-pink tinge and this colour will remain essentially the same throughout the day.

As far as the other physical and chemical properties of the Martian surface are concerned, the measurements performed at the landing sites of the two Vikings indicated a bulk density (porosity) of the surface material between 1 and 2 g cm^{-3}, an amount similar to, though somewhat higher than, that prevalent on the surface of the Moon. Other properties, such as particle size or resistance to penetration, are also similar to those found on the Moon—compare, for example, the penetration of the spacecraft's legs into the lunar and Martian ground as seen in plates 4 and 35. On Mars, more than about half of the particles of surface material on both landing sites are less than 0·1 mm in size. It is these particles which are stirred and lifted into the air by diurnal convection. As established by the x-ray fluorescence spectrometer aboard the Vikings, the chemical composition of this material indicated that some 15–30% of it by weight is silicon and 12–16% is iron, while calcium contributes 3–8%, and titanium 0·5%, other elements being present in diminishing amounts. This does not seem to make the atomic composition of the Martian crust very different from that of the Moon or, for that matter, from that of the Earth. As on the Moon or the Earth, its most abundant constituent element is oxygen and it may amount to most of the remaining percentage. However, because the instrumental limitations of the spectrometer prevented it from establishing the abundances of elements with low atomic numbers, this surmise could not be verified by actual measurements.

The Surface of Mars

A flat terrain profusely covered with boulders, both large and small—that is the dominant feature of the Martian landscape as seen from the vicinity of each landing site to the horizon some three kilometres away. What does the surface of Mars look like from a more detached vantage point? As long as terrestrial telescopes represented the only means of exploration, the observers using them recorded and mapped two classes of markings on the Martian surface: namely, 'dark' and 'bright' markings. The dark markings of lower reflectivity occupied less than one-third of the visible surface and were generally considered to form 'continents' sometimes interconnected by 'oases' or 'canals', while the bright markings of the background areas were of a reddish colour and were regarded as 'deserts'. Unlike the polar caps, the main outlines of the bright and dark markings were considered to remain relatively stable for decades or even centuries, but for the 'oases' or 'canals' this seemed much less certain and seasonal variations were similarly suspected. While the bright markings were identified (correctly) with arid desert plains and their colour ascribed to that of ferric oxides long before the Viking landers provided the final proof, the dark markings were thought to owe their lower reflectivity to moisture, or even to vegetation flourishing in the spring and watered by systems of 'canals'—natural or artificial—from the melting polar caps. Their seasonal variations were regarded as indications that at least plant life existed on this planetary neighbour of ours. These expectations were destined to total disappointment by the outcome of the Viking missions; the two colours of the Martian markings were shown to be

126 *The Realm of the Terrestrial Planets*

Plate 34. Horizontal panoramas of the Martian landscape as recorded by the cameras of Viking 2 pointed in different directions. The blunt cone seen in the centre of the lower panel is the cover of the spacecraft's nuclear power plant (cf figure 4). *Photographs by courtesy of NASA and JPL.*

Plate 35. The immediate proximity of the landing site of Viking 2 as televised on 4 September 1976 with its high-resolution camera capable of resolving details less than a millimetre in size on the Martian surface. The centre of the field of view is only about 1·4 metres away from the camera. Note the fine-grained deposit (stirred from the ground by the retro-rockets prior to landing) that has settled in the concave part of the footpad and also the weathered surfaces of the surrounding rocks which testify to the power of aeolic erosion on the Martian surface. *Photograph by courtesy of NASA and JPL.*

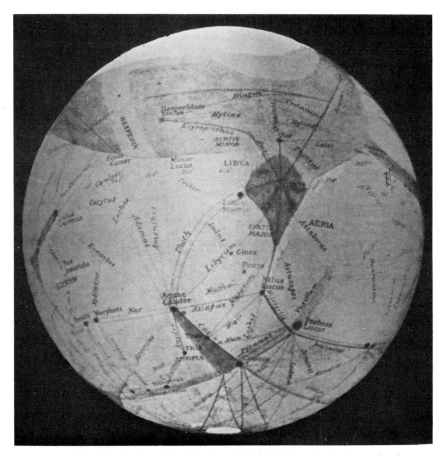

Plate 36. A view of the Martian globe based on visual telescopic observations made in the early years of the twentieth century. Except for isolated features, such as the polar caps or Syrtis Major, it bears little or no resemblance to what was discovered by recent spacecraft operating in the planet's proximity. The main reason for the difference is the fact that most details recorded on drawings were below the optical resolving power of the telescopes employed.

due to those of the limonites and basalts coexisting in an arid environment.

Many drawings and photographs of all kinds of such features abound in the popular astronomical literature of older vintage, and two examples are reproduced in plates 36 and 37 to enable us to place the contributions made since 1960 in proper historical perspective. Earlier students of areography have seen, and were bewildered by, so many different kinds of markings. Bewildered is certainly the right word; for topographic details reported by different observers were often so much at variance with each other, and so close to (and often below) the limits of

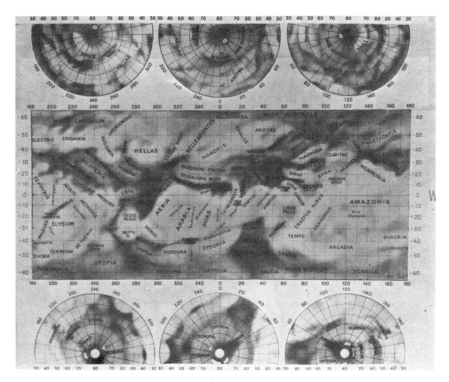

Plate 37. A somewhat more up-to-date map of the Martian surface recorded in the mid-twentieth century, and representing the extent of our knowledge of the Martian topography before the advent of the space age.

optical resolution of their telescopes, that doubts were repeatedly expressed by the more sober students of the subject about the reality of these markings. As we have learned in the past few years, such doubts were indeed well founded.

The first contribution to a more accurate study of the surface came from the establishment of radar contact with Mars in 1963. This was much more difficult than in the case of the Moon. The first difficulty was due to the much greater distance of Mars from the Earth. Unlike the intensity of the light of a planet (which diminishes with the inverse square of the distance), the strength of a radar echo, that is, the reflection of a signal sent out from the Earth, attenuates with the inverse fourth power of the distance of the reflecting body. The second difficulty was due to the larger size of Mars (twice the dimensions of the Moon) and its much higher rate of spin, both of which widen the frequency profile of the returning echoes and thus make their detection more difficult.

But, in spite of these problems, records of positive echoes of increasing precision have been obtained since 1963 by many groups of investigators. The observed strength of the returning echoes proved to be consistent with the assumption that the Martian landscape is covered by dry rocks whose dielectric properties are close to those of the iron oxides. In addition, the use of range-Doppler techniques enabled radar to be used as a sensitive recording method to transcribe the shape of the Martian surface. We have already mentioned that the shape of Mars is essentially spheroidal and flattened at the poles by its axial rotation, but by how much does the actual surface of Mars depart from that of a rotational spheroid? Unlike some of the lunar orbiters, the Martian orbiters did not carry any laser altimeters which could track the shape of the profile of an intersection of the Martian surface with the plane of their orbits. So far it has been necessary to use Earth-bound radar, and such work, carried out since 1963, has brought many surprises. First, it revealed that the surface of Mars is considerably warped (i.e. it departs from that which would be expected for a state of hydrostatic equilibrium) on a scale much greater than the centimetre to metre scale at which it is very rough. It exhibits differences in elevation from a smooth surface (corresponding to Martian 'sea level') as large as 12–16 km over long slopes covering wide areas. Moreover, no obvious correlation seems to exist between different types of surface markings and their elevations; both high and low levels occur in both bright and dark areas. To give an example, Syrtis Major, the largest and most conspicuous dark marking on the Martian globe (and one visible through relatively small telescopes), constitutes a giant incline some 1000 km in extent and rising by more than 10 km in elevation from its (astronomical) western to eastern edge. Similar differences were found to exist also elsewhere on Mars.

If we compare such departures from the form of equilibrium on Mars with those encountered on the Earth, it is true that if our oceans were removed from the surface and the latter were scanned by a radar beam from outer space, level differences greater than 10–12 km would be found between the abyssal plains of the ocean floors and the high continental plateaux of Central Asia. But the Earth is twice as large as Mars, and a level difference of 10–12 km on Mars would correspond proportionally to a terrestrial difference of 20–25 km. A difference in elevation of this size would exceed that between the top of Mount Everest and the bottom of the Marianna Trench in the Pacific Ocean. We are, therefore, inadvertently led to the conclusion that erosion has levelled off the Martian surface much less than it did on the Earth, and that the role of the principal eroding agents—air and water—has been much smaller on Mars not only now, but also throughout its astronomical past.

That this is indeed so is attested to not only by the greater proportional large-scale warping of the Martian surface, but also by its smaller-scale structure on the 1–100 km scale, as disclosed by the views of the Martian landscape televised by Mariners 4, 6, 7 and 9 (see table 6) which paved the way for the Vikings. Until the advent of space probes, limitations of astronomical optics and distortion from the atmosphere prevented telescopic resolution from the Earth of details smaller than 200–300 km in size on the Martian surface. The television cameras aboard the Mars-bound Mariners of the 1960s enabled us to increase this resolution by a factor of 100 at first and then by more than 1000 for Mariner 9 in 1971. In doing so, they opened a window for us into another world.

Craters on Mars

The television pictures relayed by the Mariners took many minutes to reach the Earth and revealed a stark arid landscape, mountainous in places, and (see plate 32) pockmarked everywhere with formations similar in size, shape, and other characteristics to the *craters* familiar to us on the Moon or Mercury (see Chapters 3 and 4). Some of the craters on Mars turned out to be hundreds of kilometres across and three to four kilometres deep, and many seemed to possess 'central mountains' similar to those appearing on the lunar surface.

The origin of most of these formations is the same as on Mercury or the Moon, both mini-planets being unprotected by any significant atmosphere and exposed to any vicissitudes which they may encounter in space. The surface of Mars (which is surrounded by an atmosphere, albeit one too tenuous to provide much mechanical protection) represents another cumulative scoreboard of the celestial target practice for meteors and meteorites, or stray asteroids or comets, whose orbits happen to be on a collision course with Mars. Because the Martian orbit does not lie very far from the inside rim of the ring of the asteroids (see Chapter 6), its surface may have experienced more hits by bodies coming from there than have either the Moon or Mercury, both of which are tucked in the gravitationally better protected inner precincts of the solar system.

For these reasons, that the face of Mars should be heavily pockmarked with impacts should have been obvious to astronomers familiar with the particulate contents of interplanetary space long before the advent of spacecraft. Of those who did realise it, that old veteran Ernst Öpik deserves an especially honourable mention. Unfortunately he was ahead of his time; and the majority of his contemporaries continued to be mesmerized by the 'dark markings' on Mars, in the belief that the alleged 'canals' were Martian watering places and harbingers of life. The existence of 'canals' has already been disproved by the Mariners and the century-old illusion of their existence must have been due to a combination of surface roughness and shadows cast by the surface relief, the details of which were below the limits of optical resolution of terrestrial telescopes. The disappearance of 'canals' under the more searching view of the optical eyes aboard the spacecraft, which could inspect the surface of Mars from a closer proximity, should caution future optimistic observers about over-interpreting their data and fishing for information inside the limits of resolution of their instruments. (It goes without saying, of course, that the same restraints should also apply to the current investigators of the data provided by the space probes.)

The principal contribution of the fly-by's of Mariners 4, 6 and 7 which reconnoitred Mars in 1965 and 1969 was the discovery that the Martian surface essentially represents an impact-mutilated landscape. The evidence supplied by the Mariners also disclosed some differences between the Martian landscape and that on Mercury or the Moon. A comparison of plate 32 with, for example, plates 25 and 28–29 shows that the Martian surface does not appear to be *saturated* with impacts to the same extent as the surfaces of Mercury and the Moon. In fact, the density of impact formations per unit area of the surface of Mars and their relatively different states of preservation are different from those on Mercury or the Moon.

The flux of impinging particles of all sizes and the duration of the exposure of the Martian surface should, of course, be no less than those suffered by its smaller planetary sisters; if anything, Mars had to absorb more cosmic impacts than either Mercury or the Moon. If, in spite of this, its surface appears today to be less densely cratered, the reason can only be the higher perishability of impact formations on Mars than on Mercury or the Moon, where much less happens that is not activated by external influences. This circumstance also makes it more difficult to reconstruct a stratigraphic history of Mars from the available photogeological records with the same assurance with which we can do so with the Moon or Mercury. In particular—and in contrast to the Moon—the testimonials from the earliest aeon of Martian history may already have been largely obliterated.

The cause of the higher perishability of the Martian surface features is no doubt the fact that, unlike the Moon or the planet Mercury, Mars is surrounded by a real atmosphere. This atmosphere is tenuous in comparison with our own (the average air pressure on the Martian ground equals that found on Earth at an altitude of 95 km) and unable to protect the Martian surface from impacts of anything larger than micrometeorites, but it is sufficient to produce other significant effects. The most important of these is *aeolic erosion* caused by the transport of dust by winds. Dust particles larger than a few dozen micrometres in size are lifted by winds to travel some distance, impinge on obstacles in their way, and bounce back into the atmosphere to be carried a step further. Particles smaller than (say) 10 micrometres are likely to remain in suspension much longer and can be lifted by convection currents to a height of many kilometres where they remain for great distances before eventually falling to the ground.

A dust storm of hurricane, or rather jet-stream, strength raged on Mars for several weeks following the approach of Mariner 9 in 1971 and virtually obliterated the visibility of the Martian surface from space during that time. The winds swept the planet's surface with a speed of 100–200 km h^{-1} from the same direction for weeks and their sand-blasting effect can be visualized by a glance at a Mariner 9 photograph reproduced in plate 38 and taken after the great storm subsided. As the wind blew past the

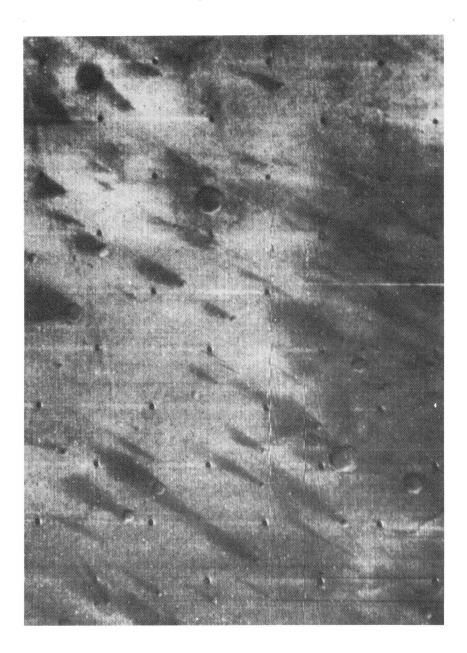

Plate 38. Another view of the Martian surface as televised by Mariner 9. The photograph shows the 'dark tails' associated with impact craters on the surface and aligned in the same direction as a result of strong winds acting on the surface material. As the wind blows past such obstacles, its velocity increases on the leeward side of the craters and thus removes the brighter dust (limonite) and exposes the darker terrain underneath. *Photograph by courtesy of NASA and JPL.*

134 *The Realm of the Terrestrial Planets*

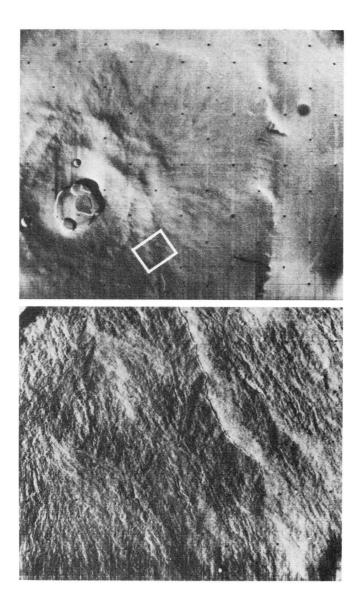

Plate 39. Two views of the cone of the Martian Mons Olympus—thought to be a volcanic caldera towering some 25 kilometres above the surrounding landscape—as televised by the cameras of Mariner 9 on 7 January 1972. The upper panel shows the summit of the main cone, while the white rectangle denotes the location of a section of the slopes shown in higher resolution in the lower panel. The complete lack of impact cratering of this area testifies to the geologically young age (not more than about 10 million years) of the whole formation. *Photographs by courtesy of NASA and JPL.*

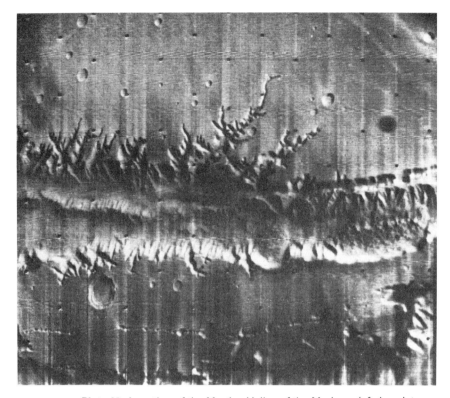

Plate 40. A section of the Martian Valley of the Mariners (cf also plate 41), as televised by Mariner 9 on 12 January 1972 from an altitude close to 2000 kilometres and covering an area 380 × 480 kilometres in size. This particularly vast chasm (in the proximity of Tithonius Lake, 460 kilometres south of the Martian equator) is 70–80 kilometres wide and several kilometres deep. It is accompanied by a whole system of tributary canyons eroded from the adjacent plateaux. In the solar system this type of formation is apparently unique to Mars. *Photograph by courtesy of NASA and JPL.*

craters, its speed increased on the craters' downwind side and removed bright dust particles from the ground to leave dark tails. It should be mentioned that the action of wind-driven particles and its significance for Martian topography was predicted by the late American astronomer Dean B McLaughlin in 1947. Although not all his views expressed at that time anticipated the subsequent findings of the Mariners and Vikings, his work on aeolic erosion and general volcanism on Mars was ahead of its time and should be remembered in this connection.

Among the most impressive features of the Martian surface are enormous formations—strongly resembling *volcanoes*—hundreds of kilometres in diameter and higher than the tallest peaks on the Earth. The youngest (best preserved) and largest is the so-called Mons Olympus. It is

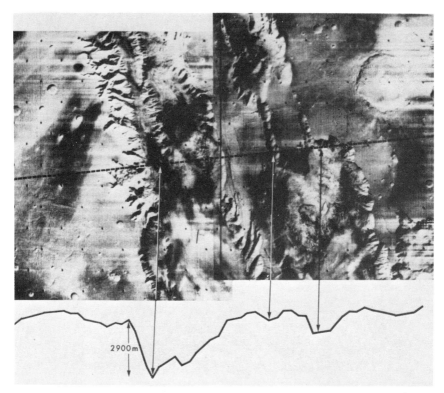

Plate 41. A section of a chasm in the Tithonius Lake region on Mars (see also colour plates 6 (a) and (b)), featuring a part of it which is 120 kilometres wide and nearly 3 kilometres deep, as recorded on this mosaic of photographs televised by Mariner 9 in January 1972. American scientists reckon that the closest similar features on Earth are the great rift valleys of Africa that run the length of the continent and that formed the Dead Sea, Red Sea, Lake Victoria and Lake Tanganyika. The Martian formation is more than twice as deep and six times as wide as the Grand Canyon in Arizona, USA. Depth measurements, indicated by the vertical profile below, were made along the dotted line by an ultraviolet spectrometer aboard the orbiting spacecraft. Two shallower rifts are indicated by the arrows to the right. *Photograph by courtesy of NASA and JPL.*

about 600 km across and its cone rises 25 km above the surrounding terrain (see plate 39). Near the summit we can see a complex of formations reminiscent of terrestrial calderas (i.e. collapse features which offer vents for lava). The slopes of Mons Olympus exhibit a rough radial texture of lobes which appear to be cratered only a little (plate 39 bottom panel), a fact from which we surmise a relatively young age of the entire formation—a few dozen million years, perhaps, but scarcely much more. There are at least three other formations of a similar type to Mons Olympus, and a number of smaller, allegedly volcanic domes about 100 km across can also be identified

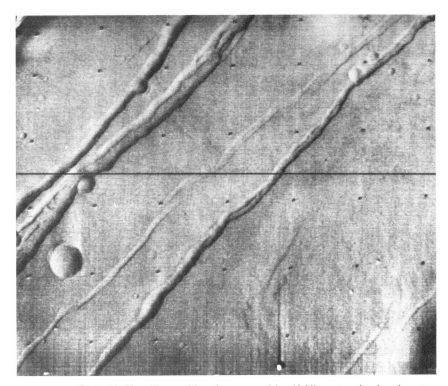

Plate 42. The rilles on Mars in an area 34 × 43 kilometres in size close to the Mare Sirenum as televised by Mariner 9 on 7 January 1972 from an altitude of 1730 kilometres above the surface. Note the striking resemblance of the rilles (common elsewhere on Mars) to similar formations on the Moon reproduced in plate 43. *Photograph by courtesy of NASA and JPL.*

on the Mariner pictures of the Martian surface. However, problems still remain with their interpretation. In order to appreciate their gravity, it is sufficient to look at the high-resolution photograph of the slopes of Mons Olympus as reproduced in plate 39 (bottom panel) and to note the complete lack of impact cratering discernible at that resolution. A terrain so free from the effects of cosmic bombardment must, astronomically speaking, be of a very recent date, and not more than (say) 10 million years could have elapsed since lava streamed down its slopes for the last time to obliterate the pre-existing sculpture. But if so, where are the gases (mainly water vapour) which should have escaped in large quantities from the crater of Mons Olympus during so recent an active phase? And, as testified by the Viking 2 seismometer, how could the Martian globe have become seismically so quiet so soon afterwards?

Another remarkable discovery of the Mariner 9 mission in 1971 was the existence on the Martian surface of *sinuous channels* and *canyons*

138 The Realm of the Terrestrial Planets

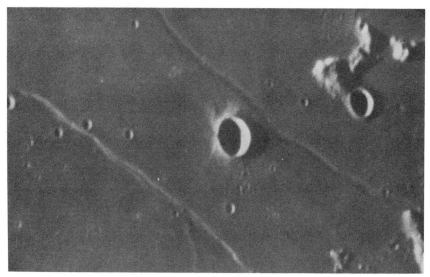

Plate 43. The crater Cauchy on the Moon, flanked by a rille and a cleft ('Cauchy's hyperbolae'), as photographed with the 43 inch reflector of the Observatoire du Pic-du-Midi (lower panel) in the course of the Manchester Lunar Programme, and by the Apollo 8 mission in orbit around the Moon on 24 December 1968 (upper panel). Note the striking similarity between these lunar features and those of Mars as shown in the previous plate. The origin of these features on the Moon has nothing whatever to do with water.

totally unrelated to the 'canals' reported by the earlier terrestrial observers. The most spectacular of these canyons, or system of canyons, has been given the name of Valles Marineris ('Valley of the Mariners') and it extends just south of the Martian equator between longitudes 45°E and 90°W. This entire system is about 2700 km long, up to 500 km wide in places, and its floor is several kilometres below the level of the surrounding landscape. One of its branches—called Tithonius Chasma—is about 75 km wide and several kilometres deep (plates 40 and 41). It may have been created by withdrawal of sub-surface material when the surface of Mars was fractured at some unspecified time, or possibly by erosion from water that may once have run over the Martian surface. Another smaller, sinuous channel is shown in plate 42. Its similarity with the lunar formations shown in plate 43 is striking, a fact which should caution us against jumping to conclusions about erosion caused by flowing water, for whatever may be the merit of such a suggestion as regards Mars, it certainly would not apply to the Moon.

When we consider these prototype formations of alleged volcanoes or canyons on Mars and compare them with their possible terrestrial homologues, the most striking fact is their truly gigantic size. Although Mars is only half as large as our own planet, everything there seems to be on a larger scale. While the summit of the Martian Mons Olympus towers 25 000 m above its surroundings, Mauna Loa, the tallest volcano and highest mountain on the Earth, rises only 10 000 m above the surrounding floor of the Pacific. If proportions were preserved, we should expect to find on the Earth volcanoes 50 km high! And what is true of the volcanoes is equally true of the canyons: while the Grand Canyon fills only a small part of the territory of the State of Arizona, the Valley of the Mariners on Mars—if transplanted on the United States—would span the North American continent almost from coast to coast!

The Interior of Mars

Could the large formations have been eroded by water on a planet whose surface today is almost bone-dry, and where a low atmospheric pressure prevents the flow of any liquid on the exposed surface? And worse: could the *interior* of Mars be hot enough to maintain large pockets of molten lava to feed its volcanoes when other, and equally eloquent, indications point out the fact that the Martian crust must be, on the whole, very much more rigid (and therefore cooler) than the crust of our own planet? Just as the orbiting spacecraft have failed to discover 'canals' on Mars, so also have they failed to discover any mountain chains which could—as on the Earth—have been raised by the *folding* of the crust.

On Mars, as on Mercury or the Moon, there is no sign of any horizontal motions of its surface layers. Indeed, with its ability to support large-scale altitude differences of more than 10 km, and with its lack of folding, the Martian crust must be endowed with a rigidity far greater than that of the terrestrial lithosphere or asthenosphere; it could scarcely possess it without being cooler. The reason is in the heat balance of a globe of given size. Since Mars is only one-half of the Earth in size and contains only a little more than 10 per cent of its mass, it could not have generated and stored nearly as much heat as our own planet. Most of the external manifestations observed on the surface appear to support this view.

Furthermore, a three-axial seismometer aboard Viking 2 (that on Viking 1 failed to return any results because of instrumental malfunctioning) did not record much—if any—seismic activity of the Martian globe at the present time. During the time of its operation the same instrument would have recorded many hundreds of seismic events on the Earth, and several dozens if it had been placed on the Moon; but hardly any were recorded on Mars. By its mass (and, presumably, internal heat supply) Mars should be intermediate in seismic activity between that of the Earth and the Moon. Why it is not is again a question to which we do not know the answer.

The apparent lack of seismic activity on Mars (at least, at present) should caution us once more against identifying formations like Mons Olympus with terrestrial volcanoes. Their superficial resemblance is indeed striking, much greater than anything we have found on the Moon. It is also possible that the large size of the Martian formations is due to the lower gravity on Mars (3·7, as compared with 9·8 m s^{-2} on the Earth) which would make it easier for internal forces to raise larger craters on Mars. But are these alleged volcanoes all dead now, and no longer act as foci of seismic disturbances?

Indeed, the internal structure of the Martian globe appears to be no more complex than that of the Moon. The observed moment of inertia of Mars, as well as the extent to which its globe yields to the centrifugal force of diurnal rotation and becomes flattened at the poles, all indicate that the density concentration inside the Martian globe remains still moderate although more pronounced than inside the Moon. If the mean density of Mars is just under 4 g cm^{-3}, it is unlikely to increase above 7–8 g cm^{-3} at its centre. In particular, we are fairly sure that Mars, unlike Mercury, contains no metallic core of any appreciable size. This we surmise not only from the nature of the external gravitational field of the planet as sampled by the orbiting satellites, but mainly from the fact that, unlike the Earth or Mercury but like the Moon, *Mars does not possess any appreciable magnetic field*. Magnetometers aboard the early Mariners failed to detect any magnetic moment of the planet stronger than a ten-thousandth of that of the Earth. Subsequent missions by the Russian Mars 3 and 5 space probes did detect magnetic phenomena in the Martian neighbourhood which are consistent with the existence of a weak dipole field of this planet (the axis of which, as on the Earth, seems inclined to that of rotation by 15–20°), but the strength of such a field would not exceed 60–70 gammas on the equator—that is, 500 times less than is the case on the Earth. For a planet rotating almost as fast as the Earth, these measurements show that the metallic 'dynamo' in the Martian interior—if any—must be very small or very ineffective. If the Martian crust is of a density close to $3\cdot 3$ g cm^{-3}, while the mean density of the planet amount to $3\cdot 9$ g cm^{-3}, then there is obviously not much room for a metallic core near the centre!

For Mars to exhibit large-scale volcanism under these conditions, now or in an astronomically recent past, suitable heat pockets would have to be located below the surface in the right places. Since Mars shows no evidence of plate tectonics and, therefore, cannot generate heat by a conversion of mechanical energy arising from friction, the source of heat would have to be radiogenic. The lava disgorged from the heat pockets through the orifices of the respective volcanoes should then show an increased radioactivity which could be measured. At present we do not know whether or not this is the case—only the future can tell.

Much as still remains to be learned about alleged volcanism on Mars before any hypotheses entertained now can be placed on a more solid footing, the same is true of the origin of the Martian canyons and sinuous rilles. If erosion by water had anything to do with their present shape, their origin would have to be relegated to a time when Mars possessed much more water than it does now, and its atmosphere would have to exert a much greater pressure than it does today in order to make fluid flow over the exposed surface possible. Some areologists indeed believe that, at some time in the past, Mars was surrounded by an atmosphere which was much denser than it is today and which dissipated in the course of time. Certain cosmochemical arguments testify against much degassing of the Martian interior in the past. One of these arguments concerns the very large ratio of the isotopes of argon on Mars (^{40}Ar/^{36}Ar $=3100$) which is 10 times as large as on the Earth. The ^{40}Ar isotope is being produced by the beta-decay of radioactive ^{40}K in the crust of the planet, while ^{36}Ar comes mainly from the deeper interior. If, therefore, the present Martian atmosphere contains a much smaller proportion of ^{36}Ar than does the terrestrial atmosphere, the likeliest explanation is a surmise that most of the Martian ^{36}Ar is still embedded in the interior of the planet.

It should be mentioned, however, that a relatively large amount of gases (10^{16}–10^{17} g) could also have been at least temporarily 'brought' to Mars by occasional cometary impacts. Similar events probably happened on the Moon as well as on the Earth and other planets exposing suitable targets for cometary impacts. On the Earth, an importation of (say) 10^{17} g of volatile compounds constituting the nuclei of comets (the only part of their anatomy which matters) would have made no appreciable difference to the chemical composition of our atmosphere as a whole. On the other hand, the Moon represents too small a target to suffer many cometary impacts and, besides, whatever gas may be acquired during these events would escape in an astronomically short time. As far as size is concerned, Mars occupies a position intermediate between the Earth and the Moon: it exposes to all external impacts (including cometary impacts) a target four times as large as the Moon; and its surface gravity is twice the lunar value thus ensuring a longer retention. If the explanation of certain surface features on Mars does require a flow of much more water than is now known to be there, did it come from the comets?

Whether or not this was ever the case still remains totally hypothetical; and as long as this is so we should keep our eyes open for alternative ('dry') explanations of the observed features. The undue obsession with water has already proved fatal to the champions of Martian 'canals' in the past. And the same may yet happen to those contemporary geologists too prone to identify the Martian canyons and rilles with their 'wet' terrestrial homologues, or the formations like Mons Olympus with volcanic calderas. To these people we should like to say: 'remember the Moon!' How many calderas were identified there by the Apollo astronauts? And how many drops of water ever wetted the floors of the lunar sinuous rilles?

Last, but not least, under these circumstances could a planet like Mars have ever given rise to *life* on its surface? Whatever hopes could have been entertained in this respect in the days of our innocence, the results of the space missions culminating in the soft landings of the Vikings in 1976 have shattered these hopes beyond resurrection. The Viking landers were especially instrumented for the detection of organic compounds in the Martian soil, and the molecular gas chromatographs aboard would have registered their presence in minute amounts. No efforts were spared to clarify the issue; even soil scooped up from under an upturned boulder (protected previously from direct sunlight) was subject to appropriate analysis.

The verdict of the analysis was unequivocal. Although the chemistry of the Martian surface turned out to be unexpectedly interesting, and many processes actually observed are still awaiting explanation, it is wholly *inorganic* chemistry and its perplexities arise from photochemical interactions between the solid material of the surface and solar radiation. No trace of any organic molecules which could lead to the formation of living matter was found at either landing site. This was perhaps not unexpected, for even before the Vikings landed the results furnished by the Mariners had proved that the Martian environment was almost as inhospitable—if not inimical—to life as that of the Moon. In 1976 the perennial hopes of finding life (at any level of evolution) on Mars evaporated, just like the canals, into the thin Martian air. It is now certain that, as living—let alone intelligent —beings, we are alone in the solar system and would have to travel through space for many years with the speed of light to find the likes of our own.

Many readers may find this part of the message of the Vikings disappointing, but is that the right reaction? Is it not better to know that, as the most advanced representatives of life on our planet, we represent a phenomenon perhaps unique within dozens, and possibly hundreds, of light years? This should at least endow us with an increased degree of cosmic importance and enhance our sense of responsibility to our terrestrial environment, the only cradle of life known to us so far. Should not the true meaning of the rarity of life in the Universe be an appeal to cherish this life and protect it from destruction or irreparable harm? We, and all our fellow-inhabitants of the Earth, now have reason to consider ourselves cosmically much more precious than our predecessors had reason to hope only one generation ago. And, therefore, why should we not respect each other's right for life, liberty, and the pursuit of happiness more so than we have been doing of late? If the messages of the Martian space probes of the recent decade were interpreted in this sense, and their implications accepted, the effort spent in sending them on their long journeys would be amply justified.

Martian Satellites

There is one more fact to be noted before we conclude this brief portrait of Mars: namely, the planet is attended by two natural satellites, although human efforts have added half a dozen artificial ones in the last decade. The two natural satellites, Phobos and Deimos (see plate 44), were discovered by Asaph Hall in 1877 with the aid of the 26 inch refractor of the US Naval Observatory in Washington, and are indeed most extraordinary celestial bodies. Phobos, the inner satellite, revolves around Mars at a mean distance of 9379 km (2·767 times the equatorial radius of the planet) in a sidereal period of only 7 hours and 39 minutes, that is, much faster than the planet rotates about its axis. Deimos, the outer satellite, revolves around Mars at a

Plate 44. The natural satellites of Mars, Phobos and Deimos, photographed with the 82 inch reflector of the McDonald Observatory in Texas. *Photograph by courtesy of the late G P Kuiper.*

distance of 23 459 km (6·92 times the radius of the planet) in a little more than 30 hours and 17 minutes.

By the rapidity of its motion across the Martian sky, Phobos almost resembles an artificial satellite of the Earth. Since it revolves in the same direction and faster than the planet rotates, for an observer on the Martian surface it moves from west to east in the sky and will not remain above the horizon for longer than 3 hours and 10 minutes. In this time it will run through a major part of its phase cycle, for the period P_s of its synodic orbit (after which the satellite returns to the same given phase), given by the equation

$$\frac{1}{P_s} = \frac{1}{7^h\,39^m} - \frac{1}{686^d\,22^h}, \tag{5.4}$$

is only 12 seconds longer than its sidereal month. Moreover, the inclination of the orbital plane of Phobos to the ecliptic (i.e. the apparent orbit of the Sun) is such that, for a suitably situated observer on the Martian surface, Phobos can transit (like Mercury for us) across the disc of the Sun, whose apparent diameter in the sky at the distance of Mars subtends an angle of only 21 minutes of arc. So rapid is the apparent motion of Phobos that even its central transit across the Sun lasts less than 30 seconds!

The second satellite, Deimos, also revolves around Mars in the direction of the planet's axial rotation. But as its sidereal period exceeds the length of the Martian day by 5 hours and 40 minutes, it moves slowly (by about $2°.8$ per hour) from east to west in the Martian sky and passes through the meridian only after a time interval of 132 hours, a time during which the planet has rotated 5·36 times about its axis. As the synodic month of Deimos differs from its sidereal month of 30 hours and 17 minutes by only 3·3 minutes, between successive passages through the meridian it follows that the satellite should run through all of its phases more than four times! To a terrestrial astronomer accustomed to the more sedate behaviour of the Moon, the Martian satellites truly display a bewildering circus in the sky!

As long as ground-based telescopes remained our only sources of information, both Phobos and Deimos were very difficult to observe and nothing was known of their size and mass. It was not until the advent of spacecraft that this lack of knowledge was rapidly improved upon. The first close-up views of Phobos secured by Mariner 9 at the end of November 1971 (plate 45) revealed that this satellite was an elongated and irregular-shaped body some 20 × 23 × 28 km in size, tumbling about its centre of mass as it revolved and exposing a variable cross section to the observer. This is why its apparent brightness as seen through the telescope was found to fluctuate in a complicated manner. In addition, its surface bore ample evidence of cratering caused by external impacts (and not by volcanic action!). The largest of these craters is almost 3 km across and it is a wonder that the impact which caused it did not smash the whole satellite into many pieces! The reflectivity of the surface of Phobos proved also to be lower than that of any other

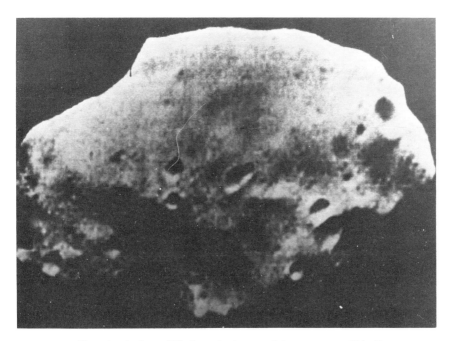

Plate 45. A view of Phobos, the larger of the two natural Martian satellites, as televised by Mariner 9 in November 1971 from a distance of 14 700 kilometres. Phobos turned out to be an irregularly shaped, distinctly elongated body 20 × 23 × 28 kilometres in size. Its surface bears ample evidence of impact cratering. That both Phobos and Deimos possess irregular shapes could have been predicted before the advent of the Mariners from the variation of their apparent brightness in the course of their axial rotations, though the cratering of their surfaces could only be discerned at a much great proximity. *Photograph by courtesy of NASA and JPL.*

known body in the solar system, lower even than that of the darkest spots on the Moon. More than 95% of sunlight incident upon the surface is absorbed and only the balance is backscattered into space to make the satellite feebly visible in optical light.

While Mariner 9 disclosed to us the size and shape of Phobos, the Viking orbiters enabled us to specify its mass. As we have mentioned already, the latter can be measured only by the effects of its attraction on any mass in its neighbourhood small enough to be affected by it; the Viking orbiters, when manoeuvred into the right position, proved to be suitable tools for this task. When the mass determined with their aid was divided by the volume of Phobos, the mean density of this satellite was found to be close to $2 \cdot 0 \, \text{g cm}^{-3}$, a value lower than that of light terrestrial rocks (such as granites) and very much lower than the mean density of the Moon. Phobos, and presumably Deimos as well, must be made of a very light material of

146 The Realm of the Terrestrial Planets

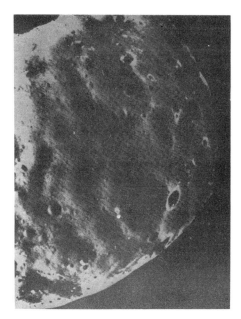 

Plate 46. Close-up views of Phobos as televised by the Viking 1 (right) and Viking 2 (left) orbiters in 1976. The view shown on the right was televised from a distance of only 480 kilometres, and that on the left from 880 kilometres. The surface of Phobos seen on the right shows parallel grooves also characteristic of certain parts of the Moon and are probably due to surface abrasion from the proto-planetary swarm. The smallest details discernible in these photographs are only 10 metres in size, a resolution attained by virtue of the fact that the spacecraft's cameras were fixed on the fast-moving objects during the exposure. *Photographs by courtesy of NASA and JPL.*

limited strength—a real wonder that it withstood in one piece the bombardment it has had to absorb. Incidentally, the Viking orbiters also secured better close-up views of the surface of Phobos at a much higher surface resolution than those procured previously by Mariner 9, and some of them are reproduced in plate 46. The resemblance between this surface (including the systems of parallel grooves, suggesting the effects of cosmic abrasion) and that of the Moon is striking and shows that both were very largely moulded by the same influences.

The second Martian satellite, Deimos, is a much fainter object, obviously smaller than Phobos, and no larger than 6–8 km in size. It too is a mere chip of rock moving lazily in space, a chip so small that an observer on the Martian surface could not discern its angular shape with the naked eye. It would appear to him only as a starlet in much the same way as artificial satellites appear to us on the Earth.

Apart from its two natural satellites, Mars has recently acquired half a dozen artificial satellites which will keep revolving around it, in different types of orbits, for an indefinite time to come. The principal characteristics of these satellites are listed in table 8. Those with orbital periods of 24 hours and 37 minutes (Mars 5, or Viking 1 and 2) are synchronous satellites revolving around Mars in highly eccentric orbits in one Martian sidereal day, a fact which enables them to hover above the landing sites of their companions on the Martian surface. In other words, the lander components of the combined missions always have their orbiting companions directly overhead, albeit at very different altitudes, to facilitate data transmission. Around the time of its periareum passage (i.e. its closest approach to Mars, bringing it to within 1500 km or less of the surface) the orbiter receives information by radio from the systems operating on the lander. The information is then transmitted to the Earth around the time of its apoareum passage some 30 000 km from the planet. With the Russian Mars 5 and 6 probes the lander–orbiter partnership could not be put into effect because of the failure of their landers.

The orbiters are not only passive transmitters of information gathered from the ground: they are spacecraft almost as elaborate as the landers and also have their own optical systems. With these they have been carrying out both an independent television surveillance of the Martian topography and a spectroscopic surveillance of its meteorology. Around the time of their closest approach to the planet, the orbiters have provided us with high-resolution photographs of the Martian surface showing its more panoramic views from greater distances (see colour plate 6(*b*) and plates 38–42).

The way in which Mars acquired its recent family of artificial satellites is obvious, but how could such unlikely bodies as Phobos and Deimos have ever been acquired? We suggest that these are probably nothing more than asteroids captured from the adjacent belt in which such particles abound. But how could their capture have been accomplished? A mere fly-by, no matter how close, would never meet the purpose, for the approaching particle would just swing around the more massive body in an open hyperbolic orbit—the faster so, the closer the approach. In order for such an event to result in a permanent capture, it would be necessary for the fly-by particle to lose a sufficient part of its kinetic energy to enable the planet's attraction to convert the open orbit of the intruder into a closed one and thus secure the new acquisition in its gravitational girdle.

This function is accomplished for artificial satellites by the action of retro-rockets; but what could have played their role with the two natural satellites? Tidal friction (cf Chapter 3) may possibly have been effective enough for this purpose with a satellite as massive as the Moon, but not for such tiny freaks as Phobos or Deimos. The only alternative seems to be a suggestion that when two asteroids from the nearby belt happened to pass close to Mars while it was still at its formative stage—while the future planet still consisted of a swarm of loose particles from which it eventually

Table 8. Natural satellites of Mars

Name	Orbital period	Mean distance (km)	Orbital eccentricity	Inclination
Phobos	$7^h39^m13^s84$	9 379	0·015	1°02
Deimos	$30^h17^m54^s92$	23 459	0·0005	1°82

Artificial satellites of Mars

Mission	Weight (kg)	Date of launch	Injection into orbit	Orbital altitude (km) at Periareum	Orbital altitude (km) at Apoareum	Orbital period	Orbital inclination to the Martian equator
Mars 2	4650	19 May 1971	27 November 1971	1376	24 960	18^h	48°9
Mars 3	4650	28 May 1971	2 December 1971	1488	124 000	11^d	48°9
Mariner 9	1031	30 May 1971	13 November 1971	1650	17 205	12^h24^m	64°28
			16 November 1971	1399†	17 040†	11^h59^m	64°37
Mars 5	?	25 July 1973	12 February 1974	1760	32 500	24^h37^m	35°
Mars 6	?	5 August 1973					
Viking 1	5125	20 August 1975	19 June 1976	1500	33 000	24^h37^m	
Viking 2	5200	10 September 1975	7 August 1976	1550	33 000	24^h37^m	

† Since 15 November 1971.

coalesced—a dissipation of kinetic energy necessary for capture could have been accomplished by collisions inside the proto-Martian cloud. The surface of Phobos (plates 45 and 46) certainly bears ample evidence of collisions with other types of smaller particles. It was possibly these collisions which slowed down the orbits of the natural satellites from heliocentric to areocentric ones, although not sufficiently so as to make them eventually crash-land on the nascent surface of the planet and become a part of its mass. In other words, Phobos and Deimos may have remained in orbit because the debris surrounding them was not massive enough to reduce their motion by collisions to sub-orbital speed. This is why they may have survived as satellites of Mars up to the present time.

6 Micro-Planets: Asteroids and Lesser Denizens of Interplanetary Space

The Moon and the terrestrial planets described so far are the most conspicuous inhabitants of the inner precincts of our solar system, but do not by any means constitute its entire population. Although they represent the principal actors of the show as far as mass and size are concerned, the number of smaller fry which still remain to be introduced on to the stage is enormous. As we shall see, their small masses do not represent a true measure of their significance. To put it another way, the Moon and the three terrestrial planets introduced so far represent only the extreme tail in the mass–frequency distribution of astronomical bodies of silicate–iron composition in the solar system. The main part of such a distribution consists of bodies much smaller than the terrestrial planets or even the Moon; their total mass will scarcely compare with that of any planet, but their numbers will turn out to be overwhelming.

Moreover, the distribution of these bodies in space is itself characteristic: while a large majority of the more significant members of this population—the asteroids—are located in the gap between the orbits of Mars and Jupiter (i.e. the outer rim of the domain inhabited by the terrestrial planets), particles of still smaller masses and of much greater numbers are virtually ubiquitous, and have infiltrated almost the entire solar system from its outermost parts to the close proximity of the Sun. The range in size and mass of these particles is truly astronomical, and so is the diversity of the ways by which they have brought themselves to the attention of their observers. In what follows we shall introduce these quaint denizens of interplanetary space by the 'clans' to which they belong—related as they all are by their chemical composition—and shall subject them to an interrogation which will add to our knowledge of the solar system as a whole.

Asteroids

The asteroids—or planetoids, as they should be more appropriately called, for they have nothing to do with the stars—occupy a significant place in the structure of the solar system by their numbers as well as by the space which they fill. Johannes Kepler, the discoverer of a true model of the planetary system at the beginning of the seventeenth century, was struck by an apparently empty wide gap of space between Mars and Jupiter. He speculated that a new planet may one day be discovered which would fill this gap. Unlike some of his other speculations, this one eventually came true, only in a somewhat different way than Kepler had imagined.

The first indication that the space between Mars and Jupiter was not devoid of all mass did not come until the beginning of the nineteenth century, almost two hundred years after the time of Kepler. On 1 January 1801 the Italian astronomer Giuseppe Piazzi of Palermo noticed that a small star of eighth apparent magnitude changed its position in the constellation of Taurus fairly rapidly in relation to the neighbouring stars in the sky, and mindful of the discovery of Uranus in 1781 (then recent history) he suspected at once that he had found another planet.

This discovery had a rather dramatic sequel. Illness soon prevented Piazzi from continuing his observations and by the time he had recovered sufficiently to return to his telescope, the constellation of Taurus with his object had disappeared in the evening twilight and could not be observed again until it re-emerged in the morning sky several months later. Where, then, should Piazzi look for his object after so long a lapse of time? The problem thus posed gave the young 'prince of mathematicians', Carl Friedrich Gauss, then 24 years old, a chance to show his mettle. By an application of the 'method of least-squares' (invented by him a few years earlier) he succeeded in extracting from Piazzi's observations sufficiently accurate orbital elements of the new body to predict its future positions. His prediction enabled the German astronomer, J G Olbers, to rediscover it in the sky on 1 January 1802, exactly one year after it was first spotted by Piazzi. This feat also had another sequel: it earned young Gauss an appointment as Director of the University Observatory in Göttingen, a post which he—a conservative soul—held for the rest of his life.

The celestial object discovered by Piazzi in 1801 proved to revolve around the Sun in a mildly eccentric orbit (characterized by $e = 0.08$) inclined by only $10°.6$ to the plane of the ecliptic, and with a period of 4·60 years, its position being almost exactly halfway between Mars and Jupiter. This celestial object was given the name Ceres. However, Ceres did not remain the only known celestial body between Mars and Jupiter for long. In March 1802 Olbers discovered another such object—called by him Pallas—which proved to revolve around the Sun with almost the same period (4·61 years) as Ceres. It did so in a much more eccentric orbit ($e = 0.24$) inclined by 35° to the plane of the ecliptic; but it approached Ceres closely at one point and this gave its existence away. From this close coincidence Olbers conjectured that Ceres and Pallas were originally parts of the same parent body—an assumption held for a long time and not completely abandoned even at present. While looking for possible additional cosmic splinters that might pass near the alleged common point of intersection, in 1804 K L Harding discovered a new 'asteroid' (Juno). In 1807 Olbers detected another (Vesta) which did not, however, approach the other three at any point of its orbit.

After 1807 almost 40 years elapsed before a new celestial object of this class (Astraea) was discovered in 1845. From that time the number of new discoveries began to grow very rapidly, especially since the advent of astrophotography with wide-angle optical systems in the latter part of the nineteenth century. By the year 1900, over 500 asteroids had their orbits around the Sun well determined—a number which was at least quadrupled by 1950. The number of those now known to exist within the reach of modern Schmidt photographic telescopes probably exceeds 100 000; those still fainter are truly too numerous to be described and catalogued.

By the size of their orbits (whenever known) the asteroids almost fill the entire space between Mars and Jupiter, but not uniformly so (see figure 9 for their distribution). In fact, the distribution of the semi-major

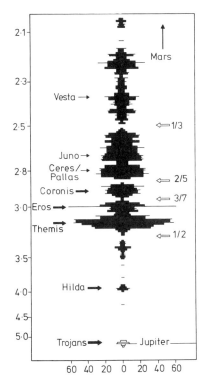

Figure 9. A frequency distribution of the dimensions of asteroidal orbits in the solar system. Vertical axis: semi-major axis of the orbit (in astronomical units); horizontal axis: number of orbits of the respective size.

axes of their orbits between 2 and 5 AU is highly uneven and exhibits a conspicuous clustering into groups of objects whose orbits are similar not only in size, but also in shape and orientation. Ceres and Pallas constitute the nucleus of one such group. Another is formed by the so-called 'Trojans' (asteroids bearing the names of the Homeric heroes of the Trojan War), which revolve around the Sun in almost circular orbits, in the same period and at the same distance as Jupiter. In fact, for well known reasons, the Trojans cluster around the vertices of equilateral triangles which their locus forms with the Sun and Jupiter†. Their number is now known to exceed 1000 (much greater than that of the combattants of the Trojan War), while the asteroids of the Themis group (see figure 9) are even more numerous.

Just as certain distances from the Sun appear to be favoured by the asteroids, others seem to be unpopular. These idiosyncrasies give rise to *gaps* in the frequency distribution of the semi-major axes of asteroidal orbits (figure 9), a tendency detected by K Hornstein and D Kirkwood in the middle of the nineteenth century. We know now that the location of these gaps corresponds to the existence of certain *commensurabilities* between the mean motion of the asteroid and Jupiter, the ratios 1:3, 2:5, 3:7 and 1:2 being particularly conspicuous. The reasons for these zones of avoidance are dynamical and due to the vulnerability of orbits in such gaps to planetary perturbations. A particle which happens to drift into these zones will soon have its orbit transformed by perturbations so as to gain allegiance to a more stable group of asteroids next to the respective gap. It may be added that, within each group characterized by the same orbital period and distance from the Sun, the shape and orientation of the orbit in space may exhibit a certain degree of dispersion. Groups of asteroids which fulfil all three conditions constitute the 'Hirayama families' (so called in honour of the Japanese astronomer who first pointed out their existence), many of which are known to exist in the sky.

The shapes or eccentricities of asteroidal orbits range widely between 0 and 1. Some, like those of Icarus ($e = 0.83$) and Adonis ($e = 0.78$), are of cometary, rather than planetary, magnitude and in the course of their orbits these asteroids experience a wide range of interplanetary climates. Thus while, for instance, Icarus (with a period of 1·12 years and an orbital semi-major axis a of only 1·08 AU) can approach the Sun to within half the distance of Mercury at its perihelion passage, Hidalgo (with $a = 5.71$ AU and $e = 0.65$) revolves around the Sun in a period of 14·0 years and at aphelion reaches almost to the distance of Saturn. The most distant

†The coordinates of this locus correspond to the so-called Lagrangian triangular solutions of the problem of three bodies obtained by J L Lagrange more than 100 years before the first such asteroid (Achilles) was discovered in 1905. Two triangular points exist in the plane of the Jovian orbit: one 60° ahead of Jupiter and the other 60° behind. The Trojans ahead of Jupiter appear (according to Tom Gehrels) to be at least twice as numerous as the host of stragglers following Jupiter from behind, but the reason for this asymmetry is so far unclear.

asteroid-like body known to us so far is the recently discovered Chiron ('Object Kowal' of November 1977), which revolves around the Sun once every 50·7 years in an orbit characterized by $a = 13·7$ AU and $e = 0·38$. The orbit makes its distance from the Sun vary between 8·5 AU at perihelion and 18·9 AU at aphelion, that is, between the orbits of Saturn and Uranus! By its dimensions, Chiron should belong among the largest known asteroids of the Pallas–Vesta class, but its distance from us should make it appear seldom brighter than a star of eighteenth magnitude and a large telescope would be required to spot it.

On the other hand, certain members of the asteroidal family, such as Eros, Apollo, Adonis, Hermes or Geographos, can approach the Earth in the course of their celestial peregrinations more closely than any other planetary body with the exception of the Moon. Two of them—Eros and Icarus—have already been contacted by radar on such occasions. Thus we cannot rule out the possibility that, one day, some may collide with the Earth, just as other asteroids have doubtless done so in the past. Indeed, the wrinkled faces of the Moon, Mercury and Mars bear ample witness to such events in records which remained legible for millions or hundreds of millions of years. The catastrophic nature of a collision with an asteroid is apt to defy the imagination of anyone but science fiction writers, and could be equivalent to the detonation of a million megatons of TNT.

In order not to scare the reader unduly with these gruesome prospects, we hasten to add that asteroids which can find themselves on a collision course with the Earth are (at least now) extremely rare. A large majority of known asteroids exhibit orbital eccentricities smaller than 0·25, just like the largest terrestrial planets; for one-half of these asteroids, e is less than 0·1. To a lesser extent, the same is true of the inclinations i of the asteroidal orbits to the ecliptic. For some, such as Pallas (for which $i = 34°.8$), the inclinations exceed those encountered among the larger planets, and no single asteroid (unlike comets!) has so far been found to follow a retrograde orbit (for which $i > 90°$). All known asteroids revolve around the Sun in the same direction as the other planets and only a few have orbital inclinations which exceed the value of 17° exhibited by Pluto (see Chapter 4).

Asteroids: Dimensions, Mass and Rotation

In Chapter 3 we mentioned that whenever we wish to study any celestial body inaccessible to direct approach, all the information we can hope to obtain must reach us across the intervening gap of space through two different channels: through gravitational attraction, and as light. And what was true for the Moon is certainly true for the asteroids. The dimensions of the asteroids are so small that their gravitational attraction cannot provide any valuable information on their mass. To give an example: Ceres, by far the largest of all the asteroids, exhibits to the telescopic observer an apparent disc only 1″.1 across—even at the time of its closest approach to us—corresponding to a mean diameter of close to 800 km (i.e. about one-quarter of that of the Moon). If the mean density of Ceres were the same as that of our satellite, its mass should be equal to 2% of that of the Moon or 0·02% of that of the Earth—a cosmically negligible amount! Therefore, the only way we can learn something about the physical (as distinct from kinematic) properties of Ceres and other asteroids is to analyse their *light*.

It should be clear to everyone that the light we receive from bodies as small as the asteroids is not indigenous, and also that they cannot possess any kind of an atmosphere. Like the Moon and the terrestrial planets, the asteroids can only be seen because they are illuminated by the Sun and because a small fraction of the incident sunlight is scattered by their surfaces in all directions, including that of the observer. It is this light which we can observe and measure with our instruments. Its amount, as well as its quality, can provide certain information on the *dimensions* and *shape* of the reflecting body.

The spectral reflectivity (broadly speaking, colour) of the asteroidal surfaces indicates that these bodies can, in general, be divided into two groups: the smaller group, approximately 20% of the total, consists of asteroids whose reflectance spectra simulate those of the stony-iron meteorites known to us from the collections in our museums and laboratories; the larger group, consisting of at least 80% of all the asteroids, exhibits spectra which resemble those of the so-called carbonaceous chondritic meteorites. From the chemical point of view, the latter represent the most primitive type of material (i.e. the least heated and the least metamorphosed) and contain appreciable traces of water and other volatile compounds. Moreover, the isotopic ratios (of oxygen, for instance) of their chondrules are different from those encountered in any other known material, and may be reminiscent of their pre-stellar past antedating the formation of the solar system. For example, the Trojan asteroids appear to be of this type and belong among the darkest celestial bodies revolving around the Sun. The Martian satellites Phobos and Deimos (cf Chapter 5) probably belong to the same class as well. As was mentioned on p 145, these are so dark that only about 5% of incident sunlight is backscattered by their rough surfaces, the rest being absorbed to produce heat.

Once we have established the reflectivity (or 'albedo') of asteroidal surfaces from their spectral characteristics, we can use the brightness of the asteroids, measured at a known distance, to ascertain the dimensions of those which are beyond the limits of telescopic resolution. All of them have been found to be considerably smaller than Ceres: Pallas appears to be no more than 500 km across; Vesta, 400 km; and Juno, about 200 km. Of those discovered later, only Hygeia (300 km) and Eumeia (270 km) or Psyche (240 km) exceed Juno in size, and a few others, such as Hebe, Iris or Metis, may approach it. Some dozens of other asteroids attain a size of the order of 100 km, while those with dimensions of the order of 10 km already number in their hundreds. Eros, around 25 km in size, belongs to this latter group as does Phobos, an asteroid captured by Mars; Deimos, of similar origin and about 12 km in size, is somewhat smaller. Asteroids like Icarus, Apollo, Hermes or Adonis—to mention only a few of those that have paid close calls on the Earth in recent decades—are probably only a few kilometres in size. Asteroids petering out into the subkilometre range are too numerous to be counted.

From known size and estimated density, the masses of the individual asteroids can be deduced to a fair degree of accuracy. The average density of asteroidal material cannot be far from $3\cdot0$ g cm^{-3} ($3\cdot5$ g cm^{-3} for typical stony-iron, $2\cdot6$ g cm^{-3} for carbonaceous chondritic meteorites). The density of $2\cdot0 \pm 0\cdot1$ g cm^{-3} recently established for Phobos (see p 145) renders it a member of asteroids of the chondritic type. In the light of these figures, and with the known size distribution of asteroidal bodies, it has transpired that, in spite of the enormity of the number of solid particles revolving around the Sun in the asteroidal belt, their total mass cannot exceed much more than three times that of Ceres alone—that is, about 3×10^{24} g of material (L Kresák 1977), which represents only $0\cdot05\%$ of the mass of the Earth or 6% of that of the Moon. More than one-half of this material is made up by the masses of Ceres, Pallas, and Vesta. The entire Trojan host numbering over 1000 does not account for more than 3–4% of the total asteroidal mass, and the Earth-crossing asteroids of the Icarus type do not contribute more than $0\cdot0001\%$. If, as some would still have it, the entire asteroidal population in the solar system constitutes the debris of a single hypothetical planet destroyed by some unspecified catastrophe, then either the debris left over represents a very small fraction of their initial mass, or this planet must have been much less massive than the Moon, let alone the Earth.

Another fundamental contribution made by photometric studies to the exploration of the world of micro-planets concerns their *axial rotation*. These studies, carried out now for quite a number of years with adequate precision, have disclosed that the apparent brightness of asteroids varies not only with the inverse square of their distance from us, but oscillates also (with clock-like regularity) in periods of less than one day lasting, in general, from a few to several hours. For solid bodies of fixed shape the periodic light variations can be ascribed only to varying cross sections exposed by such bodies to an external observer at different times. The amplitudes of the observed light changes are the measures of their departures from spherical form.

Under what conditions can a celestial (i.e. self-gravitating) body retain a spherical form? Each celestial body, whether solid or fluid, is compelled by nature to assume the form of minimum potential energy (which, for self-gravitating bodies, is a sphere) whenever the internal hydrostatic pressure is sufficient to overcome any force opposing it. Inside solid bodies the action of hydrostatic pressure is opposed by molecular forces of the cohesion of solid state. The latter represent short-range microscopic forces whose strength is independent of the total mass of the respective sample. In contrast, hydrostatic pressure depends on the total amount of self-attracting mass; if the latter increases, a situation is bound to be reached at which no solid substance will be able to withstand the ever-tightening embrace of self-gravity and its material will be crushed to a state of minimum potential energy by the weight of overlying matter. For silicate materials, whose behaviour we are mainly concerned with, this critical pressure is of the order of 10 kilobars.

How large must a self-gravitating celestial body be before the laws of nature compel it to assume spherical form regardless of composition? For fluid bodies, any mass will do, since fluids do not possess any significant internal strength of their own. For cold solid bodies of silicate composition, R Wildt (1962) found the critical radius to be a little larger than 300 km, an amount much smaller than that of the Moon (1738 km), let alone of any terrestrial planet. As was mentioned on p 47, the hydrostatic pressure of the Moon exceeds the 10 kilobar limit throughout most of its interior. This is why the shape of the Moon is essentially spherical and why it always appears in the sky as a circular disc (or a part thereof).

The asteroid Ceres, with a radius close to 400 km—in excess of Wildt's limit—can still remain essentially spherical. Its light changes of small amplitude, indicative of a nine-hour period of axial rotation, are no doubt due primarily to an unequal distribution of bright and dark regions over its surface. On the other hand, Pallas and Vesta possess dimensions well below the critical radius. Their more pronounced light changes with periods of a little over 10 hours may be due to the deviations of their surfaces from spherical form as well as to an unequal distribution of bright and dark regions. The same is true to an even greater extent of all the other asteroids which are smaller than their three large relations. As shown in plates 45 and 46, the televised close-ups of Phobos leave no room for doubt about its irregular shape. The same is true of the well known asteroid Eros—a frequent close visitor from the asteroidal community—whose occultation of the star κ Geminorum on 23 January 1975 (observed from different places on the terrestrial surface) disclosed it to be an elongated solid slab of dimensions close to $9 \times 19 \times 30$ km, in agreement with previous impressions of direct observations made with large telescopes at its earlier close approaches to the Earth in 1930 and 1952.

If bodies like Eros or Phobos originated as irregular-shaped solids, their minute dimensions should have enabled them to retain this form indefinitely. Apart from subsequent collisions with other asteroids or meteorites which could have disfigured them still further, their present geometrical characteristics should be directly related to those obtaining at their origin. But how were these acquired to begin with? One explanation offered frequently, and going back in fact to Olbers at the beginning of the nineteenth century, is the *fragmentation* of such bodies by mutual *collisions*.

Theoretical investigations of the more recent past, particularly by S L Piotrowski (1953) and E J Öpik (1963), have shown that the observed mass–frequency distributions of the asteroidal population can be made compatible with this hypothesis, a conclusion reiterated more recently by L Kresák (1977). Other arguments connected with the motions of irregular-shaped asteroids about their centres of mass have, however, been advanced (Z Kopal 1970) casting some doubts on such an interpretation. These arguments concern the fact that *the observed light changes of irregular-shaped asteroids are singly periodic* and do not exhibit secondary periodicities or beat phenomena of any kind.

In order to explain the bearing of this fact on the problem at hand, let us pause to realize that irregular-shaped debris produced by fragmentation of larger parental masses should tumble in space about their own centre of mass—just like a free-wheeling 'spinning top'—with an arbitrary distribution of rotational momenta about three axes favouring no particular direction. The motion of a spinning top can be described in terms of a three-axial rotation with periods which (for irregular-shaped bodies) are, in general, not commensurable but of comparable length. If, however, the asteroids were rotating in this manner, their light changes as seen by a distant observer should show evidence of multiple periodicity. This does not seem to be the case. In fact, the light changes of all asteroids observed so far represent singly periodic phenomena, suggesting rotation about a *single* axis with a direction *fixed* in space. But this is at variance with the view that the bodies originated from a collision between two planets (or by some other kind of similar catastrophic event) which would endow the splinters with random motions about their centres of mass.

Would it be possible for a planetary body to be destroyed in such a way that its remnants would rotate about only one axis? Such a situation could be possible if, for instance, the disruptive force were not a random collision, but the tidal action of the Sun—in other words, if a new-born planet happened to wander so close to the Sun (within its 'Roche limit') that it was torn to pieces by the Sun's attraction. This kind of mechanism has often been advanced in the past to account for the origin of the rings surrounding the planets Saturn and Uranus. Is it possible that the material now in the asteroidal belt represents depleted remnants of a similar ring which once encircled the Sun, and that the Hornstein–Kirkwood gaps represent phenomena analogous to the Cassini division and other details observed in Saturn's rings?

Such a possibility cannot yet be ruled out, but it is not very likely. In order to locate the present asteroidal belt within the Roche limit of the Sun, the solar surface would have to extend beyond the present orbit of the Earth. This may once have been the case, but could any planet have wandered so close to it? And besides, would any Roche limit exist under the circumstances? Its existence can be proved mathematically if the central body is homogeneous or nearly so, and such a condition may be fulfilled for

Uranus and Saturn. On the other hand, the internal structure of the proto-Sun contracting towards the Main Sequence would be far from homogeneous. Whether or not any limit exists within which a planetary body could be torn asunder by the tidal action of the contracting Sun is still an open question, but the answer is unlikely to be in the affirmative. It should be pointed out that the asteroidal belt, at least in its present form, is not as flat as are the rings of Saturn. Far from it: while the planetary rings are almost ideally flat (of thicknesses measured in dozens or, at most, hundreds of metres), the inclinations of the asteroidal orbits to the invariable plane of the solar system show a dispersion of the order of 10°. A dispersion of this size could not have been brought about by planetary perturbations within the age of the solar system.

This latter point is also very important in another connection. In order to bring it out, let us refer back to the beginning of the present chapter and the discovery of asteroids. If we also recall the contents of the preceding chapters dealing with the Moon, Mercury, and Mars, an assertion can be made that the existence of asteroid-like bodies somewhere in the solar system would be known to us even if Piazzi and his successors had failed to discover any, for the surfaces of the larger bodies like the Moon, Mercury, and Mars have been acting as the fly-catchers of the lesser fry in space. The catch of cosmic flies which at one time or another found themselves on a collision course with these planets and ended their careers by impacts on their surfaces is evident to posterity in the form of impact craters (plates 11–14, 28–29 and 32). In other words, the scars left behind by impacts constitute a sufficient proof of the existence of impinging bodies, the size of the scars evidencing the *calibre* of the impacta. In addition, the *form* of the impact scars can also give us at least some information about the *direction* from which the impinging body descended onto the surface. Approximately circular craters usually result from head-on impacts, while elongated formations are likely to be caused by grazing collisions.

What kind of evidence is offered by the larger planets in testimony on the type of bombardment which their surfaces have undergone in the course of their long astronomical past? This testimony is unequivocal: the bombardment has been *omnidirectional*, with impinging bodies arriving from directions distributed essentially at random. The polar regions of the Moon, as well as those of Mercury or Mars, seem to have suffered as many head-on or grazing impacts as did the equatorial regions. But if, prior to the catastrophe which brought the cosmic careers of the impinging bodies to an abrupt and untimely end, these bodies were in heliocentric orbits inclined only a little to the ecliptic, then *the majority of head-on asteroidal impacts should have occurred in the equatorial regions of the planetary surfaces* while *grazing impacts should be largely concentrated near the poles*.

This expectation requires one qualification: it would be true only if the axes of rotation of the respective planets themselves are normal to the ecliptic. This is indeed very nearly the case for the Moon or Mercury, but not for Mars whose axis of rotation is inclined to the ecliptic by 23°.5. Its orientation in space oscillates between ±23°.5 in the course of the (solar) precessional cycle which, long as it is (of the order of 10^5 years) in comparison with the precessional cycle of the Earth, is very short in comparison with the time during which the Martian surface has been exposed to external impacts. Therefore, the possible directionality of external impacts on the Martian globe could have been largely 'blurred' by its precessional motion.

As regards the Moon, its rotational axis is inclined to the ecliptic by only a little more than 5°, and the defocusing effect of its precession should be slight. But we cannot rule out the possibility that many impacts evidenced by the stony sculpture of the lunar surface were caused by bodies which, prior to impact, were not in heliocentric orbits, but were dynamical members of the Earth–Moon system, for such bodies could collide with the Moon again from any direction. However, this could *not* have been the case for Mercury. Its axis of rotation is virtually normal to its orbital plane (inclined by 7° to the ecliptic), and any astronomical body on a collision course with Mercury must have been in a heliocentric orbit prior to impact. The absence of any obvious concentration of elongated impact craters near the Mercurian poles testifies, therefore, that the asteroidal bodies which produced large impact formations visible on the surface of this planet were impinging from all directions—or, in other words, that *the impacting asteroids were moving in orbits arbitrarily inclined to the ecliptic*.

This is an important conclusion. If true it would imply that, in the first few hundred million years of the existence of the solar system (in the course of which the terrestrial planets absorbed a major part of their celestial bombardment), the inner precincts of this system were full of asteroids which described all types of orbits (as comets still do now) and which were gradually swept up by the planetary 'fly-catchers'. This would mean that, in the early days of its existence, our solar system—like a youthful galaxy— filled an essentially globular volume of space, and acquired its present, highly flattened form only in the course of time. Today, all planetary bodies ranging in size from the major and minor planets down to interplanetary dust are concentrated by the continuing and relentless action of the laws of celestial mechanics in the close proximity of the 'invariable plane' of the system; only our cometary population (analogous to Population II stars† in the Galaxy) still follows its same old ways.

All this leads us to conjecture that the asteroids, now largely limited to a belt between Mars and Jupiter, were once much more numerous and ubiquitous denizens of interplanetary space. In fact, their bulk may be identical with the legendary *planetesimals*, from whose coalescence the planets are in general supposed to have come into being, and of which the present particulate contents of the asteroidal belt may represent a surviving sample.

† First generation of stars born when our Galaxy was formed some 10^{10} years ago.

Meteorites and Shooting Stars

The asteroidal population of the solar system described in the preceding section peters out of sight with particles less than (say) 10–100 metres in size, for they cannot be observed by any known method. In order to recover the 'tail' of this population, we have to confine our attention to those bodies or particles which come close enough to become visible to us through the effects of their interaction with our planet and, in particular, with that part of it which borders on outer space: namely, our atmosphere.

Have you ever seen a meteor flash across the sky, and did you remember to make a wish? The wish was probably immaterial, for yourself as well as for the celestial body which may have prompted it. But the observation itself should be of interest, especially if you stop to think what it implies. Do you know that the fleeting luminous phenomenon which attracted your attention may have taken place more than 100 km away at an altitude of 60–80 km, and that its cause was a cosmic particle, weighing no more than one gram, which entered the upper atmosphere with a velocity of dozens of kilometres per second? The reason why so small a mass can give rise to luminous phenomena visible at such distances is the relatively high kinetic energy of the cosmic intruder. A particle weighing one gram and moving with a velocity of (say) 40 km s^{-1} possesses a kinetic energy equal to $4 \cdot 5 \times 10^{12}$ ergs, an amount equivalent to that of an 1800 lb sports car racing at 100 miles an hour. The atmospheric resistance which decelerates the cosmic particle is equivalent to a force which could bring our racing car to a standstill in a few seconds—no wonder the air in front gets heated to incandescence in the process!

In fact, most meteors of comparable kinetic energies do not last longer than a second or two, and spend themselves at altitudes between 80 and 120 km in fleeting flashes of light due partly to the glare from the cosmic intruder heated beyond the melting point of its material, and partly to the glow of the air cushion compressed in front of it. The temperature of the latter is high enough to bring about a dissociation and ionization of this gas and gives rise to a transient column of conducting gas in the wake of the meteorite which acts as a metallic mirror to reflect radar waves. Radar observations of such conducting wakes can enable us to track meteors, or showers of meteors, also in the daytime sky.

The depth of penetration of a meteorite into the atmosphere increases, of course, with its mass. The descent of meteorites weighing kilograms through denser layers of air may be accompanied by dazzling displays of light which last several seconds and which can turn night into day (plate 47). Some meteorites of this size (or what is left of them after heat ablation) may actually hit the ground as solids—and eventually wind up as new samples in the meteoritic collections of our museums. A few times in a century, the Earth intercepts meteorites which penetrate its atmosphere with residual masses of hundreds or thousands of kilograms, and the spec-

Asteroids and Lesser Denizens of Interplanetary Space 163

Plate 47. A photograph of one of the brightest fireballs, recorded accidentally on 23 September 1923 by Josef Klepešta at the Ondřejov Observatory in Czechoslovakia while taking a four-hour exposure of the Andromeda Nebula on the right. The fireball, which flared up repeatedly while crossing the field of view of the Observatory's 8 inch astrograph, was many times brighter than a full moon, and, for a few seconds of its flight, converted night into day. *Photograph by courtesy of the late Josef Klepešta.*

tacular effects (including seismic effects) of their impacts attract worldwide attention. The last such fall occurred in the Sikhote Alin Mountains of Eastern Siberia in February 1947, when a whole swarm of metallic meteorites of total mass in excess of 100 tons collided with our planet. Against bodies of this size our atmosphere (which consumes most of the small particles) ceases to provide any protection. The same is of course true of bodies still larger, such as small asteroids 1–10 km in size and with masses of the order of 10^{10}–10^{13} tons. These impinge with unaltered cosmic velocities which, for parabolic encounters, may attain 72 km s^{-1}, and with kinetic energies 10–100 times as large as the explosive power of an equivalent weight of TNT.

In such cases, the sudden conversion (within microseconds) of a large part of the kinetic energy into heat is more than sufficient to vaporize the entire mass of the intruder which explodes like a bubble of hot gas. The explosion leaves behind it a surface scar in the form of an impact crater, though the scar itself may be largely devoid of any traces of the actual material of the cosmic visitor which was blown away in the holocaust at the time of impact. Phenomena of this type, as unimaginable as they are violent, do not occur only on the Earth, but also on all the other terrestrial planets exposed to cosmic bombardment (their effects on the Moon, Mercury, and Mars have already been described in earlier chapters of this book). On the Earth, they occur only once in millions of years and yet, in spite of the perishability of their traces due to erosion by air and water, several dozens of impact craters have been identified by geologists in different parts of the world—testimonies to the fact that the days of the bombardment of planetary surfaces by cosmic projectiles of a calibre heavy enough to produce impact craters tens of kilometres across are not yet completely over, and that the supply of asteroids with Earth-crossing orbits is not yet exhausted.

The average amount of meteoritic material swept up by the Earth on its journey through space has been estimated to be approximately 400 tons a day, but most of it is irretrievably lost on arrival through evaporation in the atmosphere or through infall in the oceans. Only a minute fraction of this material can be recovered and identified. Those pieces of it that have found their way into our laboratories, and have been examined by suitable methods of analysis, were of tremendous value far outweighing their actual weight in rubies or gold.

As far as their chemical composition is concerned, perhaps the best known are the iron meteorites, the bulk of which is formed by an iron–nickel alloy with some admixture of cobalt. Needless to say, purely metallic meteorites are very much more resistant to exposure in the terrestrial environment, as well as much easier to identify (through their magnetic properties) than any other type of meteoritic material. For these reasons, their known finds greatly exaggerate their real cosmic abundance. Meteoritic irons also show evidence of peculiar crystalline structure (Widmannstätten figures!), suggestive of such an extremely slow rate of cooling at the time of crystallization as though this cooling took place in the interior of a body of appreciable astronomical dimensions.

A large majority of known meteoritic finds belong to the class of stony-iron meteorites of essentially silicate composition with different (small) admixtures of calcium and oxidized iron. As was mentioned on p 157, the reflecting properties of this type of material match fairly well the reflectance of about 20% of the asteroids, and a likely presumption is that these stony-iron meteorites resulted from the fragmentation of this type of asteroid. By far the largest fraction (80%) of the asteroids are, however, probable progenitors of the chondritic meteorites (so called because of characteristic crystalline enclosures within their matrix). One class of these—the carbonaceous chondrites—represents some of the most exciting pieces of solid matter that have ever come to the attention of the cosmochemist, although their perishability does not make them as numerous on the shelves of our museums as they probably are in space. From their relatively high contents of volatile elements (carbon) and compounds (water) they could not have been subjected to much heat since the time of their solidification; they probably represent the oldest types of solid matter that have survived the formation of the solar system, and may even contain in their chondrules vague echoes of their pre-solar past, so strengthening our surmise that asteroids of this type represent real planetesimals from which larger planetary bodies had grown in the course of time, and whose subsequent thermal differentiation led to the internal structure of the planets as we know it today.

Of greatest interest are the *ages* of the meteorites now in our hands, measured from the time when these bodies solidified. The methods by which we can establish these ages have already been described in Chapter 3 in connection with the chronology of the lunar surface. For a large majority of known meteorites, and especially for carbonaceous chondrites, the 'solidification' ages have been found to cluster around 4·6 billion years, an age close to that established for the lunar soil ('fines') (p 87). So close a coincidence is not accidental, but strengthens our conviction that this age represents a very important milestone in the history of the solar system—very probably the time of its origin.

The solidification ages of meteoritic material should be distinguished—as on the Moon—from the 'exposure' ages which can be determined from the accumulation in the material of nuclides originating from the interaction of the surfaces of meteorites with solar or galactic cosmic rays. Such exposure ages have been found by laboratory analysis to range from millions to hundreds of millions of years—a time much longer than that which has elapsed since the solidification of the respective bodies. The difference between the two can be explained if, before the time indicated by the exposure ages, the meteorites were shielded from the effects of cosmic rays by cometary ices in which they may have been embedded. For stony meteorites, we could also explain this as a result of the gradual disintegration of the crust of an originally larger body by space erosion, or again through fragmentation by mutual collisions.

The principal region of meteoritic material in the solar system is usually identified with the asteroidal belt between Mars and Jupiter at a distance of 3–5 AU from the Sun. If this is so, however, the time for a particle to spiral down from the asteroidal belt to the distance of the Earth by the cumulative effect of planetary perturbations alone would take several hundred million years, a time long in comparison with the observed exposure ages of most meteorites. Those with exposure ages of the order of 10^6–10^7 years could have traversed this distance only if they 'hitched a ride' on the nuclei of comets, having been swept up by them during their passage through the asteroidal belt and tossed away again nearer to the Sun.

It is probable that a large majority of the meteorites from the asteroidal belt which are being picked up by the Earth have indeed been imported into the inner precincts of the solar system by comets. Meteorites loosened by a gradual disintegration of cometary heads are bound to follow in the wake of their carriers at least for some time in the form of *meteoritic swarms* which are gradually depleted by planetary perturbations. Relatively young swarms are apt to be more compact, and the meteor showers observed at the time of the intersection (or close approach) of their trajectories with the terrestrial orbit are likely to be intensive, but of short duration.

Meteoritic material revolving in a swarm is sometimes concentrated in knots. An encounter between the Earth and one of these knots gives rise to short-lasting, magnificent displays of light remembered for decades by those who witness them. Such events have been staged by the Leonids of 1833 or, more recently, by the Draconids† of 1933 or 1946, which follow in the wake of certain comets (comet Tempel for the Leonids, or Giacobini–Zinner for the Draconids). Other well known showers are spread almost evenly along the entire orbit of the swarm, such as the dependable Perseids (associated with the comet of Tuttle–Swift, 1862 III), which provide celestial fireworks during several summer nights around the middle of August each year. Very often, meteor swarms survive long after the parent comet has completely disintegrated. An example of this was the comet Biela, last seen in 1852, which in 1872 became the progenitor of a meteor shower of the still-active Andromedids.

Not all meteors intercepted by the Earth on its perpetual journey around the Sun belong to specific showers. In general, meteor showers cannot retain their cosmic identity much longer than the comets that gave rise to them; they disperse in the interplanetary substrate to become *sporadic* meteors. Although this sporadic reservoir of meteoritic material is vastly greater than that forming distinct showers, the latter exist locally in

†Each meteor shower is traditionally given the name of a constellation in which its 'radiant' (i.e. a region in the sky from which the meteor shower seems to emanate) happens to be situated.

greater concentrations, and encounters with them can produce more spectacular, albeit isolated, events.

But one fact should be stressed: all meteors whose orbits have been determined with adequate precision—whether shower or sporadic—revolve around the Sun in *closed* elliptical orbits; *none* has been found to approach us along a distinct hyperbola. This means that *all meteors which we encounter belong to the solar system*; none reaches us directly from interstellar space. An acquisition of interstellar debris which could be described as meteoritic is possible in principle (and such meteors could possess ages in excess of 4·6 billion years), but no single piece of such a meteorite has come into our hands. And even if those meteors which only flash overhead in the sky were originally interstellar, they have been thoroughly domesticated in the solar system before being picked up eventually by the Earth.

Interplanetary Dust: Zodiacal Light and Gegenschein

The meteors and meteorites encountered by the Earth in space, and described briefly in the preceding section, do not represent the final stage of fragmentation of solid particles inside the solar system. A continuation of this process by mutual collisions should, in due course, result in the formation of grains of ever diminishing size (from centimetre to millimetre and submillimetre range) until so fine a dust is formed that the pressure of sunlight alone is sufficient to blow it away (as should happen to grains smaller than 0·1 μm in size). That dust so fine actually exists in the solar system has recently been proved by the fine etching of lunar crystalline rocks resulting from its infall (cf plate 21).

However, even before the manned missions to the Moon, the existence of dust in the inner precincts of the solar system was known to us from the phenomenon of the so-called *zodiacal light*—a cone of feeble light in the sky surrounding the Sun and extending from it in the plane of the ecliptic along the constellations of the zodiac (see plate 48). Indeed, spacecraft operating outside our atmosphere have recently shown that the band of zodiacal light extends along the ecliptic around the whole sky. It decreases in brightness with angular distance from the Sun and increases slightly again (*gegenschein*) for a few degrees near the anti-solar point (i.e. in the direction of the Sun to Earth radius vector projected beyond the Earth).

What is the source of this feebly luminous phenomenon? The fact that the zodiacal light is symmetrically distributed around the Sun, and that its plane of symmetry is the ecliptic rather than the invariable plane of the solar system (deviating by 1°43′ from the ecliptic), suggests that the zodiacal light is nothing more than *sunlight* scattered on solid particles pervading interplanetary space; an increase in its intensity near the anti-solar point is caused by the backscattering of sunlight from that direction.

Plate 48. The zodiacal cloud photographed at Chacaltaya in the Bolivian Andes (5200 metres above sea level) by D E Blackwell on 2 August 1958. The star trails in the sky background (produced by the rotation of the Earth) indicate the duration of the exposure. *Photograph by courtesy of D E Blackwell.*

Furthermore, the fact that the intensity of this scattered sunlight diminishes away from the Sun more rapidly than with the inverse square of the distance indicates that the number of scattering particles is greatest in close proximity to the Sun, and diminishes steadily with the inverse third power of the distance from the Sun. This consequence of the photometry of zodiacal light was subsequently confirmed by actual counts of the dust grains responsible for the scattering, using the particle counters aboard different types of space probes travelling in different directions.

We mentioned earlier that the process of fragmentation of solid particles in interplanetary space by mutual collisions of the respective particles—both large and small—is most effective within the asteroidal belt. We could surmise that not only meteoritic debris, but also interplanetary dust should be densest in this region. However, two spacecraft of recent origin, Pioneers 10 and 11, traversed the asteroidal belt en route to Jupiter in

1972–1973, and their particle counters, which were sensitive to impacts by solid grains 10–50 μm in size, failed to detect anything unusual. Contrary to what we know about the concentration of the coarser ingredients of the particulate contents of interplanetary matter in that region, the distribution of dust grains with masses less than 10^{-8} g appears to be smooth and relatively constant all the way between Mars and Jupiter. This indicates that fragmentation processes in the asteroidal belt may be grinding matter present there slowly, but the actual processes have not yet got very far; the origin of most particles of the zodiacal cloud is to be sought nearer to the Sun.

A satisfactory proof that the zodiacal light is produced by the scattering of sunlight on solid dust grains has been provided by the spectroscope. Extensive work on the spectrum of the zodiacal light has proved it to be identical with that of the Sun; its colour is the same and its spectral lines are reproduced without any essential change. On the other hand, the scattering of light on solid particles can be independent of frequency only if the size of the scattering particles is either large, or again small, in comparison with the wavelength of the illuminating source. Therefore, in our case, suitable candidates for the composition of the zodiacal cloud are either dust grains 1–10 μm in size (say), or elementary particles such as free electrons which also scatter light very effectively.

Since the spectral lines of the zodiacal light stand out as sharply as those of the illuminating source, and since their profiles are not 'washed out' by the Doppler shifts that would result from high-velocity random motions characteristic of free electrons, then the dust grains are left as the only admissible candidates for the main constituents of the zodiacal cloud. We may add that the density of free electrons in interplanetary space (as a component of the 'solar wind') has been almost continuously monitored by interplanetary spacecraft since the 1960s and found too low to be of any photometric significance. Thus, contrary to the opinions held up to the 1950s, we now know that the zodiacal light is due almost entirely to the scattering of sunlight on solid particles in interplanetary space and that free electrons play only a negligible role.

The dust grains which constitute the bulk of the zodiacal cloud must not only *scatter* incident sunlight in a manner which is virtually independent of frequency, but also must *polarize* it to a degree which depends on the angular distance from the Sun. Such a requirement imposes further restrictions on the properties of the requisite dust grains: they should be largely between 1 and 10 μm in size, elongated rather than spherical, and their refractive index should be that of carbonaceous or silicate material. The mass of an average dust grain in the zodiacal cloud should, therefore, range from 10^{-8} to 10^{-12} g or less, an amount far too small for such particles to be detected by optical or radio methods at the distance of the Earth. Photometric and spectrometric measurements of the zodiacal light thus permit us to extend our knowledge of the particulate contents of interplanetary space to very much finer ingredients.

These studies can also disclose the reflectivity (albedo) of the particles and consequently their average individual brightness. How can it be possible, we may ask, to measure the reflectivity of micrometre-size particles hundreds of millions of kilometres away? But it is possible and can be done by a comparison between the brightnesses of the zodiacal cloud in the visible and infrared regions of the spectrum. The brightness observed in the visible is scattered sunlight, but a part of the incident light must also be absorbed and re-emitted at the prevalent temperature, a process which (according to Planck's law) should take place largely in the near-infrared region of the spectrum. The observed ratio of the scattered to absorbed–re-emitted light should then indicate its reflectivity.

Recent measurements of the intensity of the zodiacal light in the infrared region of the spectrum (where most emission is of thermal origin), when compared with those of its visible (scattered) component, have revealed that the particles of the zodiacal cloud must be very dark—darker than the darkest spots on the Moon and as dark as carbonaceous asteroids or meteorites. Less than 4% of incident sunlight appears to be scattered by these materials with its spectral composition unaltered, the rest being absorbed and used to heat the particles to temperatures which may become quite high close to the Sun. According to measurements of the infrared thermal emission of the zodiacal cloud made by J H Peterson and R G MacQueen in 1967–1968, at a distance of 10 solar radii from the Sun the temperature of a 10 μm zodiacal particle should be about 1300 K, which is close to the limit at which this particle may begin to volatilize; smaller particles could be considerably cooler, for they find it easier to emit the heat which they absorb. Indeed, micrometre-size particles—especially those made of carbon—can approach the Sun in solid state to 3·5–4 solar radii before they are destroyed by the solar furnace (and beyond which they can resolidify from hot carbon gas), a fact which may also furnish a clue to their origin.

The total mass of the zodiacal cloud has been estimated from its brightness to be of the order of 10^{19} or 10^{20} g, an amount comparable with that of a single modest-size asteroid rather than that of a planet. Dispersed as individual particles weighing no more than 10^{-8} g on average, the zodiacal cloud should give rise to an optically thin cloud or haze through which we can see to any depth. This expectation is fully borne out by another striking celestial phenomenon visible during total eclipses of the Sun, namely, the solar corona (see plate 5). Part of the light of this corona, the so-called F-corona, is again nothing more than sunlight; all its spectral lines are faithfully preserved down to a distance of less than one solar radius from the apparent limb of the Sun. As was shown by C W Allen and H C van de Hulst in 1946–1947, this component of the corona originates by diffraction of sunlight on interplanetary dust along the line of sight extending from the Sun to the Earth, and not only in the proximity of the Sun. At low angular distances from the light source, diffraction becomes much more effective than scattering for the redistribution of light, an effect illustrated by the familiar example of diffraction rings around a distant lamp seen through fog.

The fact that the zodiacal cloud is optically thin should allow us to study systematic motions within it, at different distances from the Sun, from the Doppler shifts and asymmetry of its spectral lines caused by such motions. Although the existing measurements of these phenomena are not yet conclusive, they do indicate that particles of the zodiacal cloud revolve around the Sun in the same direction as the planets and with the Keplerian angular velocity appropriate for their distance from the Sun. In other words, the zodiacal cloud does not rotate like a rigid body, but revolves around the Sun like a swarm of individual particles whose mean-free path is long in comparison with the dimensions of their orbits.

This point is interesting because it can provide information on the stability and the age of the zodiacal cloud. The particles constituting it are not acted upon by gravitational forces alone: particles smaller than (say) 0·1 μm in size would be blown away by the radiation pressure of sunlight; those 10 times as large would be influenced by forces connected with the aberration of sunlight (the Poynting–Robertson effect), giving rise to a drag which would make them spiral down eventually to the vicinity of the Sun to be destroyed there. It has been estimated that the continuous operation of this process deprives the zodiacal cloud of a mass of about 10^7 g per second, or 10^{14} g per year. Since, however, the total mass of the cloud has been estimated to be 10^{19} g, it would follow that the entire cloud should be destroyed by the Poynting–Robertson effect in something like 10^5 years—a time negligible in comparison with the age of the solar system. The cloud we see today must represent only an ephemeral phenomenon which could not exist for long without being continuously replenished.

The existence of numerous meteor swarms with cometary associations suggests that comets may be carrying solid debris into the inner precincts of the solar system. But from where? We have already learned from Pioneers 10 and 11 that the asteroidal belt contains no particular reservoir of suitable material. It is possible that some of it is being accreted from interstellar space; but if so, extended periods of 'domestication' would be required before it could cascade down to the inner parts of the solar system and bask briefly in sunlight before its eventual destruction. The fact that so much of this dust surrounds the Sun may not be the work of the Poynting–Robertson effect alone, but also of a possible local production of 'smoke' in the close proximity of the Sun by condensation of gases (mainly carbon) evaporating from its atmosphere. Other stars of lower temperatures are known to produce so much smoke that it can obscure an appreciable fraction of their own light. This smoke is eventually repelled by the star to contribute to the dust found in interstellar space—a dust which constitutes an essential ingredient of the medium from which new stars are born.

From what we know our Sun makes only a very small contribution to this kind of space pollution (far less than some of its neighbours), and galactic justice is unlikely to penalize it too severely for this activity. Astronomers pleading the case can at least point out that if stars did not 'smoke', they would be impeding, rather than assisting, the birth of the next generation of stars—a fact which constitutes the main difference between the consequences of environmental pollution in the Galaxy and on our Earth.

7 Venus: The Veiled Planet

The planet Venus, our nearest planetary neighbour in space and a glorious ornament of our evening or morning sky, has claimed man's attention for almost as long as the Moon. Its changeable position in the sky attracted the attention of Babylonian observers as early as the second millenium BC and the first (known) tables of its motion were constructed during the reign of King Ammizaduga in the first half of the seventeenth century BC (see plate 49). On account of its brightness in the sky, Venus had even earlier earned for itself an honourable position in the Pantheons of many early civilizations under the double guise of Hesperus and Phosphorus, or the evening and morning star. While the Moon was the patroness of hunting and fertility (symbolized by the Greek goddess Artemis), Venus personified the goddess of love (Aphrodite), a rank to which she seemed more than entitled because of her playful changes in appearance. Her feminine nature destined her for the role of a sister planet to the Earth and she continued to play this role under her opaque veil of clouds until about 1960 when the veil was first penetrated by man-made radar and, later, by spacecraft. And, feminine to the end, how she managed to fool us under her demure external appearance!

An opportunity to study the planet arose long after the time of King Ammizaduga and his priest–astronomers who were interested only in appearances. In 1611, at the time of the dawn of telescopic astronomy, Galileo Galilei recognized, and expressed in one of his concealing anagrams (*Mater Amorum aemulatur Cynthiae formas*) that, to a telescopic observer, Venus simulated the phases of the Moon (see plate 50). This fact had important cosmological consequences at that time for, according to Ptolemaic astronomy (in which both Venus and the Sun were supposed to revolve around the Earth), Venus was located between the Earth and the supposed orbit of the Sun. But if so, the planet could not show us more than one-half of its illuminated globe—that is, it should exhibit a crescent not exceeding first quarter. Therefore, when Galileo saw with his telescope that Venus—even though it never elongated from the Sun in the sky by more than 48°—also exhibited *gibbous* phases, it became apparent that Venus must revolve around the Sun rather than around the Earth. Thus the Ptolemaic system of the heavens was dealt its death-blow. However, the Copernican system was not enthroned by this alone, for the facts observed by Galileo were compatible not only with this system (according to which all

Plate 49. A cuneiform tablet (No. K160 of the British Museum collection) with an ephemeris of the planet Venus, compiled during the reign of the Babylonian King Ammizaduga (1646–1625 BC), tenth king of the Amurru dynasty and successor (four times removed) of the great lawgiver Hammurabi. *Photograph by courtesy of the Trustees of the British Museum in London.*

planets, including the Earth, revolved around the Sun), but also with the Tychonic one (according to which all planets except the Earth revolved around the Sun and then, together with the Sun, around the Earth). A decision between the two systems had to be based on other grounds, and these had little to do with Venus.

Subsequent telescopic observations of Venus disclosed that this second-innermost planet of the solar system revolves around the Sun in 224·70 days of terrestrial time, in a slightly eccentric orbit ($e = 0·0068$) with a semi-major axis of 108·2 million km. The orbit is inclined to the orbital plane of the Earth by 3°23′40″ which, at inferior conjunction, brings Venus to within 40·7 million km of the Earth. At such times, Venus becomes our nearest planetary neighbour, though it is still more than 100 times as far from us as the Moon. A space probe which reaches the Moon after 60–70 hours of free flight must spend at least three months on its way to Venus; even light or other electromagnetic signals sent out from the Earth need at least 140 seconds to reach it (in contrast with 1·28 seconds to the Moon). On the other hand, at the time of superior conjunction, when Venus is on the opposite side of the Sun, its distance from us increases to 258 million km, a distance covered by light in 14·3 minutes.

Venus: The Veiled Planet

Plate 50. The planet Venus at five different phases of its synodic orbit as photographed by H Camichel with the 15 inch refractor of the Observatoire du Pic-du-Midi.

Between these distances the apparent diameter of Venus, as seen from the Earth, ranges from 10 seconds of arc at superior conjunction to 64 seconds (more than one minute of arc) at the time of its closest approach. At a distance of 40·7 million km this corresponds to a globe of radius only a little less than 6100 km, which does not deviate significantly from a sphere.

Whenever it happens to be visible at dusk, Venus appears to be brighter than any other object in the sky (apart from the Sun and the Moon) and its apparent visual magnitude ranges from −3·3 to −4·2. Since Venus revolves around the Sun in an orbit interior to that of the Earth, it is bound to exhibit a whole range of phases from 'new' to 'full', just like those of Mercury or the Moon (plate 50). At the time of superior conjunction its full phase can, of course, be seen only in daylight with the aid of a telescope. As it moves from superior towards inferior conjunction in the course of its synodic year and comes closer to us, an increase in apparent diameter (see again plate 50) will at first more than make up for a diminution of the illuminated part of its disc, and the planet grows brighter. Beyond a certain phase, however, its crescent becomes so narrow that its apparent brightness begins to decrease in spite of its increasing proximity.

As a result of a combination of these effects, Venus attains its maximum brightness, not when it is nearest to us (because, at that time, its phase is almost 'new'), but about 36 days before or after inferior conjunction. At that time, its elongation from the Sun is 39° (i.e. less than the maximum elongation of 47°.5) and, when viewed through a telescope, it would appear to be like the Moon about two days before 'first quarter'. At such times, Venus is bright enough for a terrestrial object to cast a shadow in its light, and can be seen in the sky with the naked eye in full daylight (if one knows where to look for it).

Before the advent of spacecraft, the mass of Venus could be deduced only from the perturbations caused by this planet to the motions of Mercury or the Earth. The extent of the perturbations disclosed that the ratio of the mass of the Sun to that of Venus was close to 408 000. More recently, the ratio was determined with much greater accuracy by fly-by space probes, several of which have been sent out since 1961 to reconnoitre the environment of Venus at close range (table 6). As a result of a very accurate radio tracking of the motions of these spacecraft in the proximity of Venus, we know now that the ratio of the mass of the Sun ($m_\odot$) to that of Venus ($m_\venus$) is given by

$$\frac{m_\odot}{m_\venus} = 408\,520 \pm 10. \tag{6.1}$$

Accordingly, the ratio of the mass of Venus to that of the Earth is 0·815 03. In absolute units, the mass of Venus is equal to $4\cdot871 \times 10^{27}$ g, a value not very different from that of the Earth. As the radius of Venus of 6100 km deduced from visual measurements was shorter than that of the Earth by only 270 km, it was thought for a long time that Venus and the Earth are not only located next to each other in the solar system, but are almost twin sisters in their physical characteristics as well. Yet appearances are often deceptive; and this proved indeed to be true of Venus. For all the superficial similarities in mass and size, investigations over the past 20 years have disclosed that vast differences exist between the other physical properties of these two planets—differences which did not become apparent until the advent of radar and space astronomy.

Exploration by Radar; Axial Rotation

In a very real sense, the exploration of Venus commenced on 10 March 1961 when the first radar contact was established with that planet from the Jet Propulsion Laboratory of the California Institute of Technology. Signals sent out at a wavelength of 12·5 cm produced detectable echoes within a few minutes of a double-transit time. Since those days further important contributions have been made by the Lincoln and Arecibo Observatories in the United States, as well as by the Radiophysics Institute of the USSR Academy of Sciences.

The basic principles of the methods by which a study of radar echoes from planetary surfaces can contribute to our knowledge of the physics and astronomy of these bodies have already been explained in Chapter 4 and need not be repeated here. The determination of the time-lag between the outgoing and returning signals, coupled with the known velocity of propagation of electromagnetic waves through empty space, has helped us to specify the distance separating us from Venus at any time with a precision far surpassing that of any previous astronomical triangulation. Radar measurements made at different times (i.e. when the planet occupies different positions in its orbit around the Sun) have led to a determination of the orbital elements of Venus so precise as to mark an entirely new epoch in celestial astronautics. For instance, the semi-major axis of the Cytherean orbit ($0·723\,329\,860 \pm 0·000\,000\,002$ AU) is now known correctly to within two parts per billion. The length of the astronomical unit was deduced by the same method to be equal to $149\,597\,892$ km (or $499·004\,786$ seconds of light travel) within an error of only ± 5 km (or 17 microseconds of light travel). This value is more than 2000 times as precise as the best previous 'astronomical' determination of the unit by celestial triangulation and exemplifies the overwhelming progress made by use of these new methods.

The greatest and most spectacular contribution of radar astronomy to the study of Venus has been the determination of its *axial rotation*. For a planet like Mars, which exhibits distinct surface features, the duration of its day and the position of its axis can be ascertained telescopically from periodic changes in the appearance of its disc. However, Venus exhibits no distinct markings which could serve for this purpose. This fact led the early astronomers to suspect that what we see on Venus through a telescope is not the actual surface of the planet, but that the surface is concealed from view under a permanent *layer of clouds*, of almost uniform brightness, hovering high in the atmosphere. The motions of any visible surface features refer, therefore, only to the movements of these clouds and not necessarily to the solid surface hidden below their impenetrable veil.

Attempts made in the past to measure spectroscopically the motion of the surface of Venus in the line of sight (by placing a slit of the spectrograph in different positions and measuring the respective Doppler shifts) referred, as we now know, only to the motions of gas in the upper atmosphere; their only message was a suggestion that the axial rotation of this planet, if any, was likely to be slow. However, although the Cytherean atmosphere is totally opaque to optical light, it becomes fully transparent at centimetre wavelengths so that radar signals can be reflected from the actual surface and have their profiles influenced by the planetary spin.

The nature of this influence has already been explained in Chapter 4 in connection with similar effects on Mercury. For Venus, the 'radar depth' of the planetary globe, that is, the time-lag between the return of signals from its limb and a subterrestrial point, turned out to amount to one-fiftieth of a second, corresponding to a radius of the solid globe of the planet equal to 6056 ± 1 km. This value is almost 50 km shorter than the 6100 km radius established by telescopic observations. The *difference* between them is significant and *indicates the altitude of the cloud layer of Venus above the solid surface*. In order to obtain the mean density of the solid globe of this planet we should, therefore, divide its mass by the volume of a sphere with a radius of 6056 km. In this way the density is found to be equal to $5 \cdot 25$ g cm^{-3}, a value only 5% smaller than that of the Earth ($5 \cdot 52$ g cm^{-3}). The area of the solid surface of Venus is 0·9495 times smaller than that of the Earth, and the gravitational acceleration on the surface is $8 \cdot 88$ m s^{-2}; the velocity of escape from the gravitational field of the planet is $7 \cdot 32$ km s^{-1}.

An analysis of radar echoes reflected from the Cytherean surface can, in fact, yield much more than the absolute dimensions of the solid globe. A range-Doppler analysis of the frequency profiles of these echoes disclosed that Venus *rotates* in a *retrograde* direction (i.e. one opposite to its orbital motion), in a period of $243 \cdot 1 \pm 0 \cdot 1$ terrestrial days and about an axis inclined by 87°.8 to the orbital plane of the planet. This was a most unexpected result, for in the entire solar system only one other planet has a retrograde rotation—Uranus; its axis of rotation deviates from its orbital plane by only 8°. For Venus, the axis of rotation is tilted virtually 'upside down'—a truly unique case.

Even more noteworthy is the fact that a period of axial rotation of 243·16 days—almost identical with that deduced from radar data—would allow Venus to show us always the same face at each inferior conjunction! In order to prove this, let us recall that the period P_V, in which the planet returns to the same position in its orbit around the Sun, is equal to 224·70 days. The synodic period P_s, after which Venus will return to the same phase for a terrestrial observer, is given by the equation

$$\frac{1}{P_s} = \frac{1}{224\overset{d}{.}70} - \frac{1}{365\overset{d}{.}26} \qquad (6.2)$$

and is equal to 583·9 days. This period is much longer than the Cytherean 'sidereal year' of 224·70 days because, at its greater distance from the Sun, the Earth follows Venus along the ecliptic with inferior speed. Accordingly, a little more than one year and seven months of our own time will elapse on average before Venus, after overtaking us, will catch us up from behind.

On the other hand, the period of axial rotation D (the sidereal day) established from radar data is equal to 243·1 terrestrial days and is related to the mean solar day D' on Venus (equal to the time interval between two successive passages of the Sun through the meridian of that planet) which can be found from the equation†

$$\frac{1}{D'} = \frac{1}{243\overset{d}{.}16} + \frac{1}{224\overset{d}{.}70} \qquad (6.3)$$

to be equal to 116·8 days, such that

$$\frac{P_s}{D'} = \frac{583\overset{d}{.}9}{116\overset{d}{.}8} = 5. \qquad (6.4)$$

As the Earth rotates quickly and revolves slowly, the difference between the mean solar and sidereal day amounts to only 3 minutes and 56·6 seconds, the solar day being the longer of the two because the Earth rotates in the same direction as it revolves. However, as Venus rotates in the opposite direction, the Cytherean solar day of 116·8 terrestrial days is considerably shorter than the sidereal day of 243·1 terrestrial days and equal to exactly one-fifth of the Cytherean synodic year of 583·9 terrestrial days.

Further surprises arise when we examine the time interval T between successive meridional transits on Venus, not of the Sun, but of the Earth. The reciprocal of T should be equal to a sum of the reciprocals of the sidereal day on Venus and the sidereal year on Earth, and follows from the equation

$$\frac{1}{T} = \frac{1}{243\overset{d}{.}16} + \frac{1}{365\overset{d}{.}24} \qquad (6.5)$$

to be equal to 146·0 days, such that the ratio

$$\frac{P_s}{T} = \frac{583 \cdot 9}{146 \cdot 0} = 4, \qquad (6.6)$$

illustrating that T is equal to exactly one-quarter of the Cytherean synodic year. The striking coincidences between equations (6.4) and (6.6) imply

†The right-hand side of this equation consists of a sum of two terms, rather than a difference, because the rotation is retrograde.

that, between each two successive conjunctions, Venus completes *four* rotations for the terrestrial observer, but *five* rotations with respect to the Sun.

For an observer situated on Venus (above the clouds), because the planet rotates in a retrograde direction the Sun would rise in the west and reach the meridian almost one terrestrial synodic month (29·2 days) later. After another such month the Sun would set in the east and the cycle would repeat itself every 116·8 days. Since the axis of rotation of Venus is almost perpendicular to its orbital plane, and since its orbit is almost circular ($e = 0 \cdot 0068$), there should be no significant seasonal changes on Venus in the course of the year; the climate should remain the same at every latitude. The Earth is in opposition with Venus every fifth Cytherean solar day. At these times, an observer on Venus above the clouds would see the same face of the full Earth at midnight as a brilliant, dazzling celestial object of apparent visual magnitude $-5 \cdot 6$. The Earth would appear much brighter to the observer than Venus ever appears to us, and at no more than half a degree to the side of the Earth he would see the Moon as a star of first magnitude some 100 times fainter than the Earth, but still conspicuous to the naked eye.

The remarkable resonance represented by equation (6.6) between the synodic orbit of Venus and its axial rotation with respect to the Earth is certainly not accidental. It strongly suggests the existence of *tidal coupling* between the two neighbouring planets, but the specific mechanism which could lead to its establishment is still largely obscure. In Chapter 4 we showed that Mercury was tidally coupled with the Sun, but a similar coupling between Venus and the Earth—a body very much less massive—constitutes a real challenge to our understanding. The same is, of course, true of the anomalously slow axial rotation of Venus in the retrograde direction. These phenomena constitute the main, unsolved dynamical problems which we have encountered in the inner precincts of the solar system so far.

The Atmosphere of Venus

As we mentioned earlier, if the visible surface of Venus shows no recognizable surface markings of any permanence and appears to a telescopic observer as a uniformly bright disc scattering incident sunlight more effectively (up to 59%) than any other terrestrial planet, then it is virtually certain that what we actually see is not a solid surface, but a continuous layer of opaque clouds permanently surrounding the entire planet and supported by an extensive atmosphere.

Telescopic observations indicating the presence of a Cytherean atmosphere have been available for a long time. Near the time of inferior conjunction, the horns of the Cytherean crescent often extend noticeably beyond the sunrise termination, and when the planet is very close to the Sun near its superior conjunction, the cusps of the crescent have been observed to coalesce into a complete aureola surrounding the whole disc (plate 51). This phenomenon can be due only to the diffuse reflection of sunlight in a gaseous medium, the same process in fact which produces our twilight. The total amount of gas necessary to produce the observed aureola can be

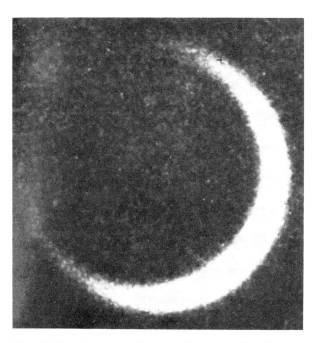

Plate 51. An almost complete aureola surrounding the apparent disc of the planet Venus near the time of its superior conjunction, and arising from diffuse reflection of sunlight in the dense Cytherean atmosphere. Photograph taken in June 1940 at the Lowell Observatory in Flagstaff, Arizona. *Reproduced by courtesy of the late J B Edson.*

estimated from the intensity of the reflected sunlight: it is equivalent to a layer of gas about one kilometre deep at terrestrial atmospheric pressure. This refers, of course, to the amount of gas above the top of the visible cloud layer; what lies below this layer cannot be disclosed by this method.

What does the upper atmosphere of Venus consist of? In 1932, the American astronomers W S Adams and Th Dunham detected in the near-infrared region of the Cytherean spectrum three complex bands due to the molecular absorption of carbon dioxide (CO_2). The measured intensity of these bands proved that the amount of CO_2 above the clouds must be equivalent to a layer not less than three kilometres deep at terrestrial atmospheric pressure. Such an amount of CO_2 is about 1000 times as large as that in the entire atmosphere of the Earth!

Apart from CO_2, which is the main constituent of the Cytherean atmosphere, the only other gas molecules detected through the specific absorption features of their infrared spectra were carbon monoxide (CO), detected by W M Sinton and V Moroz, and water vapour (H_2O), detected by R A Shorn and others from the ground, or by J Strong and his collaborators from high-altitude balloons. The amount of water detected by these observers was, however, very small and corresponded to only 10 micrometres of precipitable liquid, an amount 10 times smaller than that found by the Vikings on Mars (cf p 115) and incomparably smaller than that on the Earth. Thus the Cytherean atmosphere above the clouds turned out to be excessively dry, and the same condition was later proved to be true of its lower layers as well.

Before we descend into the depths of the lower Cytherean atmosphere concealed beneath clouds impenetrable to ordinary light, let us examine the *temperature* of Venus as deduced from the intensity of its thermal emission in different regions of the spectrum. The bulk of this emission by planetary bodies with temperatures of a few hundred degrees kelvin will (in accordance with Planck's law) peak in the near-infrared region, well beyond the limits of light visible to the human eye. When the first reliable measurements of infrared emission from Venus between wavelengths of 8 and 12 μm were performed in 1955 by E Pettit and S B Nicholson, it transpired that the mean temperature above the clouds was approximately equal to $-35\,°C$ and remained constant day and night.

In order to penetrate deeper into the Cytherean atmosphere, the thermal emission has to be measured at wavelengths long enough not only to emerge from the veil of clouds, but also to penetrate the atmosphere of the Earth. This double requirement implies that we must use wavelengths greater than one millimetre, that is, we must examine the *radio spectrum* of Venus. Exploration by this method was begun in 1956 by C Mayer and his colleagues at the Naval Research Laboratory in Washington, and its results could not have been more surprising: whereas near the top of the cloud layer the intensity of thermal emission at 0·01 mm wavelengths indicated a prevalent temperature around $-35\,°C$, the intensity at wavelengths between 8 and 9 mm indicated temperatures in excess of $+80\,°C$, rising in the centimetre range to values close to $+400\,°C$!

At these wavelengths, the atmosphere of Venus becomes effectively transparent—just like our own. Therefore, a temperature of $400\,°C$ corresponding to the intensity of the Cytherean emission in this domain of the spectrum refers undoubtedly to the solid surface of the planet. If, however, such a temperature is to be attained as a result of the 'greenhouse effect' in the Cytherean atmosphere (i.e. by intake of solar heat through visible light, the escape of which is prevented by absorption of the atmosphere's gaseous constituents), and if the principal absorber were carbon dioxide, then its amount would obviously have to be enormous. Calculations of the model-atmospheres have disclosed that, to account for a ground temperature on Venus of $400\,°C$, an atmosphere consisting predominantly of CO_2 would have to be dense enough to exert a ground pressure of close to 100 terrestrial atmospheres!

When these figures were first produced, they were received by astronomers with general disbelief. However, more information was soon to come from a completely different source which left no room for doubt and confirmed the previous findings in a resounding manner: namely, the space probes sent out to Venus to reconnoitre its environment at close range (see table 6). The first terrestrial messenger of this kind was the Russian probe Venera 1 launched in February 1961. Although it failed to reach its goal (radio contact with it was lost after a distance of 7·5 million km), it proved to be a harbinger of greater things to come. Its abortive journey paved the way for the fly-by's of the American spacecraft Mariners 2 and 5 (which paid close calls on our sister planet in December 1962 and October 1967) or the Russian Venera 2 (in February 1966). The subsequent Veneras 3–9 (between 1966 and 1972; see table 6) entered the Cytherean atmosphere and through the use of parachutes actually descended to the surface with various degrees of success.

Venera 3, the first spacecraft sent out from the Earth to land on another planet, touched down on the surface of Venus on 1 March 1966 but failed to return any information. However, Venera 4 performed some measurements on site, as did its successors to an increasing extent. Venera 9, one of the more recent spacecraft (October 1975), provided the first televised views of the surface of Venus (see plate 54). A brief account of these activities reveals that, whereas the space exploration of Mars has so far been largely accomplished by American space probes, that of Venus has been done predominantly by Russian ones. In saying so we do not, of course, wish to belittle the fine contributions to the exploration of Venus made by the American Mariners 2, 5 or 10, for the latter especially rendered a signal service to the exploration of the outer Cytherean environment. But no spacecraft of American origin has landed on the surface of Venus so far, and none is scheduled to do so in the foreseeable future.

From their latest descents through the Cytherean atmosphere, the Venera landers have revealed that, by mass, about 93% of this atmosphere is composed of carbon dioxide, with nitrogen (both N and N_2) and probably argon making up the balance. Water vapour seems to amount to no more than 0·01% of the Cytherean atmosphere above the cloud cover (not much more than 0·1% below), and oxygen still less; the corresponding percentages of the noble gases are still unknown. The scarcity of oxygen is striking and certainly presents an enigma. The carbon monoxide on Venus is presumably formed in its upper atmosphere where carbon dioxide can be dissociated by the ultraviolet light of the Sun. An inevitable by-product of this reaction should be oxygen, and since molecular oxygen is diatomic, the number of oxygen molecules should be equal to one-half the number of carbon monoxide molecules. This is indeed the ratio we find on Mars—another planet whose atmosphere consists very largely of carbon dioxide—

but on Venus oxygen seems to be at least 50 times less abundant than carbon monoxide.

Another important contribution by the Veneras to the study of the conditions on Venus has been to verify *in situ* that the air pressure on its surface does attain 95 atmospheres and that the ground temperature exceeds 460 °C—a veritable inferno which Dante, if he had known about it, would certainly have chosen as a suitable abode for the most obdurate of sinners! A medium so hot and so dense would be sufficient to kill any vestige of life. Measurements made by Venera 8 in 1972 also suggested that only about 1% of sunlight incident on Venus ever reaches the surface. In spite of the proximity of this planet to the Sun, it should be about as dark there as it is on Earth on a very foggy day and, as in a fog, objects should cast only very weak shadows.

With increasing altitude the visibility should improve, but only slowly because the layer of clouds extending between 40 and 50 km above the surface continues to obscure the view. Only at altitudes greater than 60 km can the top of the cloud cover be reached. It is at this level that anyone emerging from the Cytherean atmosphere would get a glimpse of the starry sky for the first time. The pressure drops to one atmosphere near the 50 km level and to one-tenth of an atmosphere at the top of the cloud layer where a temperature close to -35 °C prevails. Far below the deceptive calm of these rarefied heights, clouds impenetrable to visible light conceal a veritable inferno in an eternal semi-twilight which no man can ever enter and hope to return; even spacecraft find it difficult to function for any length of time.

The Veil of Clouds

What do the clouds on Venus really consist of? To determine their composition is far more difficult than to identify the gases constituting the atmosphere, because these clouds must consist of particles which are either solid or liquid. Such materials, unlike gases, do not produce spectral lines that could identify them. When the clouds surrounding Venus are observed from outside in the visible part of the spectrum, they appear to be uniformly pale and of a yellowish colour, and scatter approximately 59% of the incident sunlight. In ultraviolet light distinct features begin to emerge and some regions appear darker than others (plate 52). If these markings are caused by an uneven distribution of some material which absorbs ultraviolet radiation, then what is it?

In the absence of direct spectroscopic evidence, the only way to narrow down a list of possible candidates for this material is to study the optical properties (reflectivity, colour and polarization) of sunlight scattered by the clouds. That the light of Venus is partly polarized and that its degree of polarization varies with the phase have been known since the 1920s, when Bernard Lyot laid down the fundamentals of planetary photometry. Recent theoretical studies have shown that, in order to match the observed polarization curves, the cloud particles must be spherical rather than elongated, about one micrometre in size, and their refractive index should be close to 1·44.

This list of specifications imposes several constraints on the admissible choice of cloud constituents and eliminates candidates such as dust grains stirred from the surface, ice crystals (akin to those constituting the terrestrial cirrus clouds), or organic compounds (for instance, formaldehyde) polymerized by the action of sunlight. The spherical shape of the particles implies that they are liquid droplets rather than solid crystals. This

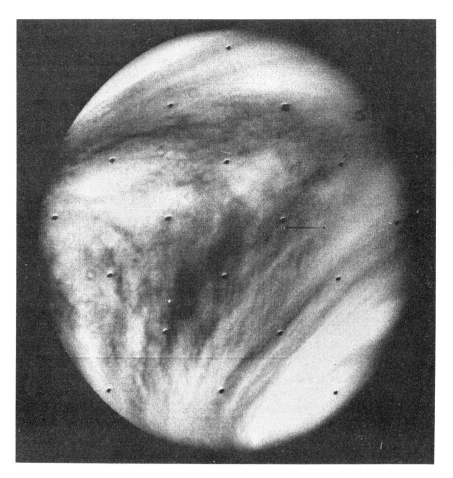

Plate 52. An ultraviolet picture of the planet Venus and its cloud structure as recorded on 9 February 1974 by the Mariner 10 spacecraft at a distance of approximately 2·8 million kilometres, four days after its closest approach to the planet. The morning termination is to the right. *Photograph by courtesy of NASA and JPL.*

condition eliminates water vapour, which would freeze at the elevation of the cloud tops into ice crystals of hexagonal rather than spherical form; in addition, the refractive index of these crystals (1·33) is too low. Most of the other inorganic substances are eliminated because their refractive index is too high.

It was only quite recently that this enigmatic situation was enlightened by W S Benedict's discovery that the high-dispersion spectra of Venus obtained by Pierre and Janine Connes by use of their powerful multiplex spectrometer indicate the presence in the Cytherean atmosphere of hydrogen chloride (HCl) and hydrogen fluoride (HF) gases. Both gases are highly corrosive and when dissolved in water yield hydrochloric and hydrofluoric acid, respectively. Their abundance is too low for them to be the principal constituents of the clouds, but they may play other, secondary roles.

Another substance which may be present in the clouds is sulphuric acid (H_2SO_4). Sulphur trioxide (SO_3), the anhydride of sulphuric acid, can be formed on Venus by photochemical reactions activated by sunlight. The spectrum of SO_3 has not been observed on Venus, for any SO_3 formed in the cooler atmospheric layers would immediately combine with any trace of water to form a haze of sulphuric acid. This hypothesis accounts reasonably well for several observed features which have baffled astronomers for some time. In particular, water solutions of more than 70% of H_2SO_4 have the proper refractive index and are liquid at $-20\,°C$. As in the terrestrial clouds consisting of water vapour, droplets begin to form at an altitude some kilometres below the cloud tops. When the pressure on Venus reaches a few atmospheres, as it does at altitudes between 30 and 40 km, a rain of sulphuric acid may begin to fall through increasingly hotter air, in which water progressively evaporates and sulphuric acid becomes more concentrated.

A rain of hot, concentrated sulphuric acid presents an intimidating prospect for the designer of a spacecraft which is to operate in such an environment. There are, however, even worse possibilities. A small amount of hydrogen fluoride (the existence of which has been spectroscopically established in the Cytherean atmosphere) would react with the sulphuric acid to form fluorosulphuric acid (HSO_3F), one of the strongest mineral acids which dissolves many metals and most rocks. A rain on Venus may, therefore, consist of droplets of the most corrosive natural fluid existing anywhere in the solar system!

The reader who may be deterred by such a prospect from going there to see for himself should, however, be encouraged not to take these current cosmochemical scenarios as gospel truth. Although the sulphuric acid hypothesis accounts satisfactorily for many optical properties of the clouds surrounding Venus, it cannot explain one important property—the yellowish colour of the planet. This colour must be produced by some substance which absorbs in the blue and ultraviolet regions of the spectrum.

Sulphuric acid does not meet this requirement; nor does any other likely substance which could be considered in this connection. Perhaps the absorber we seek to identify is a more complicated compound of sulphur with other elements that are available on Venus—or perhaps these are the particles of molecular sulphur itself (with carbon suboxide polymers representing other possible candidates).

Surface Structure and Internal Constitution

Now let us descend again through the mist or rain of sulphuric acid (or worse) to the surface of the planet and set foot on its solid ground. What is the structure of this ground, and what does it consist of? As we cannot see it from the Earth, our first source of information has again been the radar signals sent out from the Earth and bounced back from the Cytherean surface.

The strength of these returning echoes per unit area is related to the dielectric properties of the reflector and to the relative smoothness of the reflecting surface (on the scale of the wavelength of incident signals). From observations made since 1961 it transpired that the surface of Venus appears to be much smoother than that of the Moon, Mercury, or Mars, and its dielectric constant, comparable with that of dry, sandy terrestrial deserts, is indicative of a much more compact surface than that of the Moon or even Mars.

More recently, range-Doppler radar tracking (see p 178 and figure 8) of the Cytherean surface has shown that the global shape of Venus deviates locally less from that of a spheroid than does that of Mars or the Earth, and that its surface appears to be generally flat with few features more than a kilometre high. Plate 53 shows a view reconstructed from radar data of a pair of overlapping crater formations on the Cytherean surface, the largest of which is about 160 km across—a somewhat larger version of the overlapping craters Theophilus and Cyrillus on the Moon (plates 13 or 27). However, in contrast with their lunar homologues, Cytherean craters exhibit a much more shallow vertical profile with rims not more than half a kilometre high, and their floors are virtually at the same level as the surrounding landscape. Nevertheless, underneath her veil of clouds, the face of this celestial goddess of love has already disclosed to the inquisitive radar eye some traces of wrinkles (albeit not very deep) which betray her cosmic age.

As on the Moon, Mercury, or Mars, these formations, and others akin to them, are no doubt of *impact* origin and were caused by collisions between Venus and large meteorites about one kilometre in size. Against hits of this calibre even the dense and extensive Cytherean atmosphere affords little or no protection. On the other hand, the relative flatness of the terrain on a small scale (indicating a scarcity of smaller craters) can well be

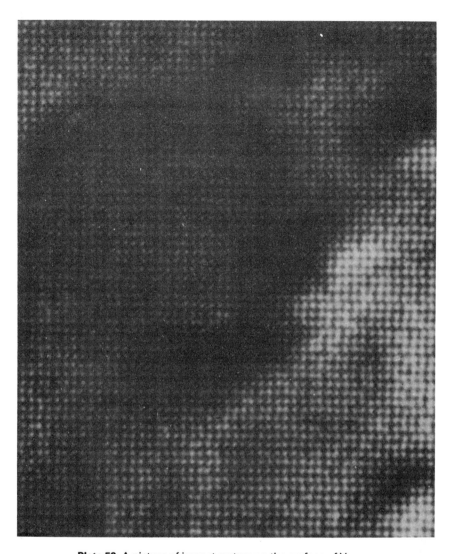

Plate 53. A picture of impact craters on the surface of Venus as reconstituted from radar which can penetrate clouds opaque to radiation at higher frequencies. The crater in view is about 160 kilometres across and its ramparts are only half a kilometre high; the floor does not appear to be much below the level of the surrounding terrain. *Photograph by courtesy of NASA and JPL.*

attributed to the atmospheric shield, for it is doubtful if meteorites smaller than (say) a few dozen metres could penetrate the atmosphere and hit the ground; even larger bodies could do so only with greatly reduced velocities.

The general flatness of the Cytherean terrain may also be due to atmospheric erosion. In the hot, dense, and corrosive atmosphere the weathering of the surface may smooth out its vertical unevenness on an astronomically short timescale, provided of course that winds of appreciable speed sweep horizontally across the Cytherean landscape. From spectroscopic observations we know that, above the layer of clouds, the Cytherean atmosphere is agitated by horizontal winds with velocities of 100–200 km h^{-1}. Below the clouds, the descending space probes registered much milder winds: the soft-landing Veneras recorded no more than gentle breezes blowing over the surface with velocities less than a few kilometres per hour.

Below the cloud layer, the motions of gas streams should be predominantly vertical and propelled by convection. This latter mechanism offers the only effective way on Venus for the transport of heat from the surface to the cloud layer. This layer may, in fact, be maintained at its elevation by updraft from the ground in the same way as the individual cumulus clouds in the terrestrial atmosphere 'sit' on top of their convection columns. To this extent, the Cytherean meteorology may be similar to that on the Earth, but one weather prediction should always be fulfilled on Venus: whatever other conditions may obtain, the weather should always remain *very hot*.

The first glimpse of the actual structure of the Cytherean surface provided on a metre scale by the Russian Venera 9 on 22 October 1975 in the close proximity of its landing site (see plate 54) shows that the surface is rough and rocky, and not dissimilar to that on Mars (colour plates 5, 6(*a*) and 7(*a*)). Should this first glimpse into a new world prove to be representative of the Cytherean surface at large, then the extent of aeolic erosion there is very limited and in agreement with the low wind velocities (2–4 km h^{-1}) recorded by the Veneras so far.

As regards the physical and chemical compositions of the boulders on the Cytherean surface as seen in plate 54, experiments performed by the Veneras indicated that their density should be close to 2·8 g cm^{-3}, a value virtually identical to that of terrestrial granites. The ground radioactivity, as inferred from the measured intensity of γ-ray emission at the landing site of Venera 8, also seems similar to that of our granites, though substantially lower (basalt-like) radioactivity was reported by Veneras 9 and 10. In these properties the surface of Venus seems to resemble the crust of the Earth and not the surface of the Moon. This single fact may represent only a slender peg on which to hang more elaborate theories on the internal structure of the Cytherean globe. From the little we know so far it is not unreasonable to suppose that the internal structure of Venus is similar to that of our Earth.

In particular, Venus is thought to possess a liquid conducting core surrounded by a mantle and a rocky crust. The size of the core and the thickness of the mantle are not known from any direct (seismic) evidence, but according to the equilibrium models constructed by R A Lyttleton in 1963, the core of total mass of 24·8% of the planet attains a pressure of

190 *The Realm of the Terrestrial Planets*

Plate 54. The first pictures from the surface of Venus, transmitted by the Venera 9 and 10 spacecraft on 22 and 25 October 1975, and showing the Cytherean landscape in the proximity of their landing sites. Venera 9 landed on a slope inclined by about 30° to the horizontal (upper panel), while the inclination of Venera 10 (lower panel) was only about 8°. The boulders with which both fields of view (separated by about 2100 kilometres) are replete are reminiscent of the surface of Mars (see colour plates 5 and 6 (*a*) and (*b*), or plates 33 (*a*) and (*b*) and 34); their sizes are likewise mostly between 0·3 and 1 metres. *Photographs by courtesy of the USSR Academy of Sciences.*

$2 \cdot 7 \times 10^{12}$ dyn cm^{-2} and a density of $11 \cdot 5$ g cm^{-3} at the centre, and extends to $50 \cdot 1\%$ of the planet's radius. The density inside the mantle of such a model diminishes from $9 \cdot 5$ g cm^{-3} at its interface with the core to $5 \cdot 6$ g cm^{-3} at the base of the crust of density $2 \cdot 9$ g cm^{-3}. Nothing can be said so far about the temperatures inside the Cytherean globe as long as we are ignorant of its internal energy sources.

One additional fact concerning the deep interior of this planet should be mentioned, and that is the apparent absence of any *magnetic field*. Measurements made by Venera 4 in 1967 indicated that if such a field exists, it must be more than 10 000 times weaker than that of the Earth (i.e. only a few gammas in strength). This essentially negative result has been fully confirmed by subsequent space probes. The effective absence of any magnetic field can, however, be understood if planetary fields are produced by the 'dynamo effect'. Of the two prerequisites of a field generated in this manner, namely, a rotating and a conducting core, Venus may possess the latter; but it generates no magnetic field because it rotates so slowly—far more so than all of its planetary sisters in the sky.

Venus: The Mystery Wrapped in an Enigma

Since the discovery of the telescope and, in particular, since the advent of space probes, we have learned a great deal about our nearest planetary neighbour. However, what we have learned so far renders Venus a more mysterious celestial body than any other planet in the inner precincts of the solar system, and its observations have raised more problems (often of a quite unexpected nature) than they answered. In order to point these out, let us say no more about the cloud layer surrounding the planetary globe in its perpetual veil, but instead focus our attention on the mysteries inside the enigma.

The first mystery concerns the rotation of the planet. What made Venus rotate so slowly, and what tilted its axis of rotation almost upside down to give rise to its retrograde rotation? The only probable mechanism would be a very close encounter with another celestial body whose gravitational attraction played havoc with Venus and altered some of its kinematic properties beyond recognition. Was this the outcome of an encounter with some long-lost planet of our solar system? Or was (a much less likely event) the disturbing body an interstellar intruder? We cannot say.

But if this, or some other, cause could be invoked to account for the slow axial rotation of Venus in the retrograde direction, its tidal coupling with the Earth—the second mystery—is much more difficult to understand. That tides are powerful enough to produce a coupling of Mercury's rotation with the Sun is understandable, at least qualitatively if not quantitatively; but a tidal coupling between Venus and the Earth is almost negligible. Moreover, both the masses and distances of these two planets must have remained essentially the same throughout the entire history of the solar system. But if so, how did their couple arise to produce the effects we now observe? Again, we cannot say.

The third mystery concerns the atmosphere of Venus—not merely the high layer of clouds (whose total mass is no doubt very small), but all the carbon dioxide contained in the lower layers. Indeed, the outstanding contribution made by recent space probes has been the realization of a great difference in the total atmospheric air masses surrounding Venus, Earth, and Mars. The atmosphere of Venus is almost 100 times more extensive than our Earth's, while that of Mars amounts to less than one-hundredth of the terrestrial air mass. It is true that as far as carbon dioxide is concerned, the atmospheres of Mars and Earth contain comparable amounts of this gas. But why does Venus contain 100 times as much?

Most of the terrestrial CO_2 is contained in solid state in the form of carbonate compounds (such as limestone) which are very common in the crust of our planet. Geochemists have estimated that the total decomposition of all these carbonates could generate an envelope of CO_2 around the Earth of some 80 atmospheres air pressure. If this were to happen, the total supply of both gaseous and fossil CO_2 on Venus and the Earth would be of comparable amounts. Indeed, the sole difference may be that, whereas most of the terrestrial CO_2 has been deposited in the solid carbonates of the Earth's crust, on Venus it has remained in gaseous form.

But what could be the reason for such a difference? One clue is probably the even greater disparity between the water contents of the two planets. On the Earth, the global amount of atmospheric water vapour gives rise to a partial pressure of one-thousandth of an atmosphere. If all the water from the seas and oceans were to evaporate, the air pressure on the Earth would increase 400 fold. On the other hand, in the Cytherean atmosphere water is at least 10 000 times scarcer than CO_2 and, therefore, its total mass should be equal to some $5 \times 10^{23} \times 10^{-4}$ g $= 5 \times 10^{19}$ g—in contrast with the $1 \cdot 4 \times 10^{24}$ g of water in both the terrestrial atmosphere and hydrosphere. Provided that the interiors of both planets are now essentially desiccated, the water on the Earth appears to be more than 30 000 times more abundant than on Venus, and the actual disparity may be even greater. The large amount of water on the Earth may explain why most of the terrestrial CO_2 has been locked up in solid carbonates, since water is an essential ingredient for their formation. But why is Venus (like Mars now) so excessively dry? And is the relatively large amount of gaseous CO_2 on Venus or Mars primordial, or was it liberated gradually by desiccation of the interior or the crust?

Suppose, for the sake of argument, that Venus initially possessed as much fossil CO_2 in its crust as did the Earth, and also a comparable amount of water. Because of its closer proximity to the Sun, Venus would receive more heat than the Earth to evaporate water from its oceans. The absorption of heat by water vapour liberated in this way would give rise to a 'greenhouse effect' which would accelerate evaporation. The surface would then become hot enough to drive out CO_2 from the solid carbonates, and thus the Cytherean atmosphere would assume its present proportion.

So far so good; but what happened to the water afterwards? How could so much of it be lost in the course of time? It is true that water vapour in the upper atmospheric layers can be dissociated by ultraviolet sunlight into hydrogen and oxygen. The former could have been removed from the gravitational field of the planet by thermal escape or by its interaction with the 'solar wind', and the latter used up for oxidation of the solid surface. Unfortunately, the efficiency of this mechanism would be quite insufficient to desiccate Venus to its present state. Of the water left on Venus today†, very little reaches the upper atmosphere where it could be dissociated. At the present rate of dissociation, Venus could not have lost any significant amount of primordial water throughout the whole history of the solar system.

Besides, the entire argument defended by some planetologists contains one fatal flaw—an implicit assumption that in the early days of the solar system the youthful Sun was as bright as it is now. Actually, as a zero-age Main Sequence star our Sun should have been approximately 40% less luminous than we see it today, and its surface temperature some 10% lower. Venus is 28% closer to the Sun than ourselves and any element of its surface receives almost twice as much heat as we do on the Earth. Even so, the early climatic conditions on Venus could not have been much warmer than they are in the terrestrial tropics today, and had there once been oceans on Venus, very little of their waters could have evaporated. Therefore, we are back where we started in our beating around this particular bush: if there is so little water and so much carbon dioxide on Venus today, the most likely reason is that the situation was the same from the beginning. But why?

Finally, another fact disclosed in 1975 by Venera 9 may force us to reconsider also some of our other conclusions. On p 182 we ascribed the high temperature encountered on the surface of Venus to a greenhouse effect produced by the powerful absorption of the plentiful CO_2 molecules in the infrared region of the spectrum. Some investigators also invoked the presence of dust (which can absorb light of all frequencies and not only in the discrete bands of the spectrum) stirred by atmospheric convection currents to increase this effect. According to their views, very little sunlight should

†If this water was really the end-product of a process of desiccation as outlined above, it should be enriched by an appreciable proportion of 'heavy water' (D_2O; in which atoms of deuterium have replaced atoms of hydrogen) which, on account of its 10% higher molecular weight, escapes less readily from the planet's gravitational field. Hypothetical, dried-up Cytherean oceans should, therefore, have left behind a greater proportion of heavy water than that contained in the terrestrial oceans, regardless of the contents of deuterium in the solar atmosphere or solar wind. No measurements of the isotopic composition of vestigial water on Venus are available so far; but if and when these are made by space probes the outcome will be awaited with interest.

reach the surface of Venus directly and its illumination by light multiply scattered by the dust should largely eliminate shadows of its relief. Now the first pictures relayed to us from the Cytherean surface in October 1975 (see plate 54) revealed that the latter is not as dimly lit as was thought and that the ubiquitous boulders in the proximity of the landing site of this spacecraft do cast distinct shadows in solar illumination. This shows that the Cytherean atmosphere must be more transparent to visible light than it would be if it contained an appreciable amount of dust. Thus the quest for the real cause of the Cytherean greenhouse effect must be renewed. With its deep atmosphere and hot, dry surface Venus continues to present us with a clouded crystal ball and ample opportunities for all kinds of further guesswork.

8 Our Earth: The Queen of the Terrestrial Planets

In approaching the conclusion of our narrative, the time is now ripe to introduce the queen of the terrestrial planets—the Earth. Our planet deserves this accolade for several reasons. Firstly, it is the *largest* and most massive of all the terrestrial planets in the solar system, and all its other physical properties are but consequences of this fact. Secondly, these properties render our Earth the only truly *multicoloured* planet of the solar system; it is the uncontested beauty-queen by virtue of the make-up of its surface (see colour plate 8). Finally, the Earth is the only planet in the solar system which, in the fullness of time, has given rise to *life*; it has eventually become the cosmic home of intelligent beings who can contemplate their station with understanding.

Vital Statistics: Motion, Size and Mass

By its position in the solar system, the Earth is situated between Venus and Mars, as though Hephaistos, smoking from terrestrial volcanoes, intended to keep Ares (Mars) apart from his golden Aphrodite. The Earth revolves around the Sun in a nearly circular orbit ($e = 0.0167$) at a mean distance of 149.598 million kilometres—a distance traversed by light in just under 500 seconds—in a period of one sidereal year of 365.256 36 . . . mean solar days, or 365 days, 6 hours, 9 minutes and 9.35 seconds. This annual revolution around the Sun is not the only motion performed by our Earth in space. It also rotates in a period of 23 hours, 56 minutes and 4.09 seconds (a time interval we call the mean solar day) in the direction of its motion and about an axis inclined to its orbital plane by 23°27′17″.6. Moreover, the direction of this axis does not remain fixed in space, but precesses slowly—like a spinning top—in a period close to 25 800 years. This motion, which is responsible for the 'precession of equinoxes' of our celestial coordinate system, is the reason why the 'tropical year' (i.e. the time interval between two successive passages of the Sun through the 'vernal equinox', marking the ascending node of intersection between the orbital plane of the Earth and its equator) lasts only 365.242 20 . . . mean solar days and is shorter than the sidereal year by

20 minutes and 23 seconds. The incommensurability between the rotation and revolution of our planet (i.e. the lengths of the sidereal year and of the mean solar day) has been the cause of endless perplexities for those endeavouring to set up a practicable calendar for civilian use.

For a long time the rate of rotation of the Earth (i.e. the length of the sidereal day) had been regarded as a master clock measuring the uniform flow of cosmic time. It is no longer sufficiently accurate for such a purpose today, for extended astronomical observations and more accurate measurements of time made by atomic clocks over shorter intervals have detected secular changes in the rate of rotation of the terrestrial globe, due partly to minor displacements of mass in the Earth's interior, and partly to secular operation of 'tidal friction' (see p 95) by which the lunar tides can transfer momentum from the Earth's rotation to the Moon's orbit. As a result of this process we know that the length of the day is secularly increasing, and that some 500 million years ago it amounted to only about 21 hours (cf p 96).

The idea that the Earth was a sphere emerged in the fifth or sixth century BC and was supported by the much-maligned Aristotle in the fourth century BC. His arguments were based on observations of eclipses (i.e. the circularity of the shadow cast by the Earth during eclipses of the Moon) and the validity of these arguments has not been weakened by the passage of time. Having decided that the Earth was a sphere, these early scientist–philosophers were naturally anxious to establish its *size* as well. It was obvious to Aristotle and his contemporaries that the Mediterranean world in which they lived must represent only a small part of the total terrestrial surface. In the third century BC, Eratosthenes of Cyrene, the great geographer of the Hellenistic times who spent most of his life in Alexandria, attempted to determine the dimensions of our planet by the triangulation of an arc of meridian between Syene (the modern Aswan) and Alexandria. By a fortunate compensation of errors inherent in his work he arrived at a result which was correct to within a few per cent. Thus the knowledge of the shape and size of the Earth has been an intellectual property of mankind for about 2300 years.

In more recent times, particularly in the seventeenth and eighteenth centuries, the dimensions of the Earth were remeasured by methods similar in principle to those initiated by Eratosthenes, but very much more precise. The equatorial radius of our planet was eventually found to be equal to 6378·20 km and its polar radius to 6356·80 km. The difference of 21·40 km between the two was due to the fact that diurnal rotation produces a centrifugal force which flattens the Earth at the poles—an effect first noted in the latter part of the seventeenth century and explained correctly by Newton.

As regards the *mass* contained within the volume of this rotational spheroid, no basis existed for its determination until the advent of Newton's theory of gravitation, which enabled the mass of a body to be related to the force exerted by its attraction. On the basis of his inverse-square law, Newton related the magnitude of the terrestrial mass $m_\oplus$ to the acceleration g of a particle falling towards it by an equation of the form

$$g = \frac{Gm_\oplus}{r_\oplus^2}, \tag{8.1}$$

where $r_\oplus$ denotes the distance of the falling particle from the Earth's centre, and G is the constant of gravitation.

The value of g could be measured reasonably well in Newton's time and the dimensions of the Earth were similarly known (since 1671) to an adequate precision. However, the absolute value of the constant G was $m_\oplus$ and, consequently G from the known values of g and $r_\oplus$. This Newton himself only guessed at it by assuming that the mean density of the Earth was between five and six times that of water—a remarkably close estimate, considering the fact that nothing was known in Newton's time about the state of the Earth's interior, and that the density of the rocks found on its surface ranged only between 2·8 and 3·3 g cm^{-3}. But once we assume a given density for the Earth, the known dimensions of its globe allow us to evaluate $m_\oplus$ and, consequently, G from the known values of g and $r_\oplus$. This Newton did, and from a value of G so obtained he went on to estimate the absolute masses of other planets, as well as that of the Sun.

The first empirical determination of G was carried out in 1772 in the Scottish Highlands by N Maskelyne, the fourth Astronomer Royal. In the nineteenth century G was redetermined more accurately by laboratory experiments using torsion balances. Today the best estimate of its value is given by $G = 6\cdot667$ cm^3 g^{-1} s^{-2} with an uncertainty of a few units in the last decimal—a relatively low precision because of the fact that gravity is such a weak force. If we combine this value with the observed gravitational acceleration $g = 982$ cm s^{-2} of free fall at a mean distance $r_\oplus = 6371$ km from the Earth's centre, the total mass $m_\oplus$ of the Earth necessary to produce the requisite acceleration turns out to be equal to $5\cdot98 \times 10^{27}$ g—a number which may seem very large. If, however, we divide this mass by the volume of the terrestrial spheroid expressed in cm^3, the mean density of the terrestrial globe proves to be equal to $\rho_m = 5\cdot52$ g cm^{-3}, the highest density for any terrestrial planet, and one well within the limits estimated by Newton three centuries ago. Ultimately, the velocity of circular motion of artificial satellites revolving around the Earth should be equal to $(Gm_\oplus/r_\oplus)^{1/2} = 7\cdot92$ km s^{-1}, while that of escape from the gravitational field of the Earth (which all spacecraft have to exceed to embark on their celestial journeys) is equal to $7\cdot92\sqrt{2} = 11\cdot2$ km s^{-1}.

Internal Structure of the Terrestrial Globe

How is the mass of the Earth distributed within its globe? The fact that the mean density of the Earth proves to be almost double that of common surface rocks makes it evident that our planet cannot be homogeneous. The first clue to its internal structure proved to be the measured value of its flattening at the poles—that is, the extent to which the terrestrial globe yields to the applied centrifugal force. This extent depends not only on the size of the Earth and the angular rate of its spin, but also on the distribution of mass throughout its interior. The difference of 22 km between the equatorial and polar semi-axes indicates that the central density of the rotating globe should be no less than three times its mean density, or about 17 g cm^{-3}. Accordingly, the density of material in the central parts of the Earth should greatly exceed that of iron (at zero pressure) and approach the specific weight of gold.

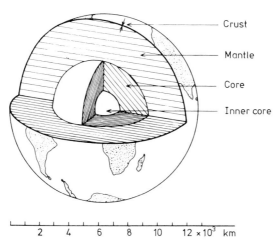

Figure 10. A schematic cross section of the interior of the Earth illustrating the principal parts of the anatomy of its structure.

More detailed information about the Earth's interior has been obtained from a study of seismic records of tectonic earthquakes that have been registered in many thousands in different parts of the world. The main result to emerge has been the realization that the Earth as a whole consists of two physically distinct parts: a *core* extending to 3470 km from the centre; and a surrounding *mantle* whose outermost shell is called the *crust* (see figure 10). By volume the core represents only about 16% of the whole Earth, the remaining 84% being occupied by the mantle. However, the mass of the core amounts to almost one-third of that of the Earth, and by its structure and composition the core is so distinct from the mantle as to constitute almost a 'planet within a planet'.

A detailed analysis of the available seismic records has disclosed a more complete picture of the Earth's interior. The crust of our planet (on which we shall have more to say in the next section) extends only down to a depth of 30–40 km, where the weight of the overlying layers gives rise to a pressure close to 9×10^9 dyn cm^{-2} or about 8900 atmospheres. At this depth the density of the crustal material increases rather suddenly from 2·8 to 3·3 g cm^{-3} and keeps increasing throughout the mantle (partly by compositional changes and partly by self-compression) up to a density close to 5·7 g cm^{-3} at its base, where the pressure has risen to over a million atmospheres. The outer parts of the mantle down to a sub-surface depth of about 400 km consist of rocks whose principal constituents are oxygen, silicon and aluminium, while at greater depths aluminium is gradually replaced by magnesium.

Some 2900 km below the surface, the composition of the Earth undergoes profound physical and chemical changes. The physical change in its state is manifested by the fact that the core of our planet fails to transmit the transversal (shear) components of seismic waves created by earthquakes sufficiently distant that their messages must reach us through the central parts of our globe. Earthquakes produce, in general, two types of waves: longitudinal (pressure) and transversal (shear) disturbances. Both can propagate through solids, but fluids are incapable of transmitting the transversal disturbances. The mantle of the Earth transmits both kinds of disturbances throughout its mass. The sudden disappearance of shear waves at depths below 2900 km discloses that the material of the core, unlike that of the overlying mantle, behaves as a fluid and must, therefore, be in a *molten* state.

Moreover, at the transition between the two regimes the density jumps abruptly from 5·7 to 9·4 g cm^{-3} at a pressure of 1·4 million atmospheres which remains continuous across the interface. So large a change in density strongly suggests that the solid mantle and fluid core differ not only in the physical state of their material, but also in chemical composition. In fact, a density of 9·4 g cm^{-3} or more can be accounted for only if we assume that the composition of the core is essentially *metallic*, its principal constituents being iron and nickel, with some admixture of sulphur or, more probably, of molten silicate oxides. Although the core occupies only 16% of the Earth's total volume, it accounts for over 31% of the mass of our planet because of its relatively high density. There are indications that, within this core (which behaves as a fluid), there exists an inner core of radius 1250 km whose material behaves again as a solid and may consist of pure metals resolidified by the enormous pressure which can exceed 3·6 million atmospheres at the Earth's centre.

As regards the chemical composition of the Earth as a whole, iron (which is stored mainly in the core) should account for almost 39% of the total mass of our planet. Oxygen, the second most abundant element by weight, contributes 27%, silicon 14%, magnesium 11%, sulphur and nickel about 2·7%, and aluminium and carbon about 1%. All the remaining elements contribute less than 2% to the total mass, of which 31·4% is stored in the core, 68·1% in the mantle, and 0·5% in the crust.

The question may immediately be asked: what can account for

this degree of differentiation of material in the terrestrial interior? Apart from a certain amount of possible differentiation that may go back to the formative processes at the time of the origin of the solar system, the principal cause is the internal *temperature*. The temperatures inside the Earth are difficult to establish by direct measurements, but borings in the crust down to a depth of 2–3 km have shown an average temperature increase of about 30 °C per kilometre (the 'geothermic degree'). At this rate, the point at which water boils (at atmospheric pressure) should be reached at a sub-surface depth of no more than 2·5 km. If surface water seeps down through occasional cracks to such a depth, it is expelled again as steam, under its own pressure, in the form of hot springs or geysers so familiar to us in many parts of the world.

If this inward rise in temperature were to continue at a similar rate to ever increasing depths, temperatures at which silicate rocks melt (about 1500 °C) would be attained at sub-surface depths of only 50 km or so, that is, depths comparable with the level of the combustion chambers of most terrestrial volcanoes. Volcanic eruptions may indeed represent nothing more than occasional outbursts of deep-seated geysers of lava and stones ejected from isolated volcanic pockets where the pressure exceeds the strength of the respective vents.

The source of heat responsible for terrestrial volcanic phenomena is likely to be mechanical, and produced by the friction of different blocks rubbing together in the outer parts of the mantle. However, such a mechanism cannot be operative much deeper in the interior and would not produce enough heat to account for the observed global value of the geothermic degree. The source of heat giving rise to it is most likely *nuclear* and due to the presence, in the Earth's mass, of long-lived radioactive elements such as uranium, thorium or the radioactive isotope of potassium. A spontaneous disintegration can not only be used to date the origin of the rocks containing these elements by the methods explained on pp 84–6, but also is accompanied by a spontaneous release of a certain amount of heat.

The amounts of these radioactive elements in the Earth's crust are, in fact, minute: terrestrial granites or basalts contain from 1 to 10 parts per million (ppm) of uranium by weight, and only 0·04 ppm of thorium. As their amounts are small, so is the amount of heat liberated by their eventual decay into radiogenic lead. However, this process has been operating continuously since the Earth was formed 4·6 billion years ago (see p 87) and the internal temperature it can give rise to depends on the balance between the generation of heat and its loss. If radiogenic heat is produced throughout the entire volume of our planet and lost through its surface and outer layers, an analysis of its balance discloses that if the mass of the Earth contained as much uranium or thorium in the interior as it does in the crust, the amount of radiogenic heat produced in 4·6 billion years, and largely contained in the interior (from which an escape by conduction constitutes a very slow process), would be sufficient to raise the internal temperature of the Earth far beyond the melting point of its material both in the core and throughout most of the mantle.

Seismic observations of shear waves throughout the mantle prove that this is not the case, for these waves cannot be transmitted through molten material. We are, therefore, led to the conclusion that radioactive elements like uranium and thorium cannot be distributed in equal amounts throughout the whole mass of the Earth, but that its crust has been enriched with them by processes which are as yet unclear. When we take into account the effects of pressure on the melting points of terrestrial materials, it becomes apparent that the temperatures prevailing at the core–mantle interface cannot exceed 5500–6000 °C, values which may remain about the same down to the centre because of the high thermal conductivity of the metallic core. In other words, the core of the Earth may be almost as hot as the surface of the Sun, and its material would radiate similarly if only we could see it. Fortunately, we are protected from its glare by the thick layer of overlying 'ash' whose surface is not too hot for us to walk upon. But if we could penetrate sufficiently far below its placid landscape, what an inferno we would find!

Needless to say, these internal temperatures are high enough to keep the metallic core in a molten state and to produce by thermal convection a self-excited dynamo responsible for the maintenance of the Earth's magnetic field. Yet whatever temperatures may now prevail in the Earth's interior, they cannot remain constant, but must change in the course of time. A warm Earth must constantly redistribute its internal heat supply by conduction to the surface where some of it escapes by radiation into space. At the same time new heat is being generated in the interior by the spontaneous disintegration of radioactive elements.

As in the stars, the main part of the thermal energy now produced in the Earth is, therefore, also of nuclear origin, though the reactions producing it are quite different. In stellar interiors, fusion reactions prevail (mainly a conversion of hydrogen into helium), and the conditions are sufficiently extreme for the rate of these reactions to be influenced by the environment. In the Earth (and other planets) all nuclear reactions are restricted to the spontaneous disintegration of the heaviest natural elements, operating at constant rates and completely oblivious to ambient conditions.

If radiogenic heat generated in this manner is indeed due to the spontaneous disintegration of elements like uranium, thorium or the radioactive isotope of potassium—the half-lives of which range between 1·3 and 13·9 billion years, that is, times of the same order of magnitude as the age of the Earth—a more detailed analysis of the terrestrial thermal balance shows that the Earth must at present still be warming up (albeit at a very slow rate), and was internally cooler in the past. If so, however, it is possible that the new-born Earth possessed **no metallic core**, and that its iron and nickel were extracted only gradually from its rocks by rising internal temperatures. At the time of the formation of the Earth, its internal structure need not have been characterized by the same degree of differentiation as it is today, and nor does its present profile, as shown in figure 10, need to represent the final stage in its evolution.

And, *mutatis mutandis,* the same is probably true of the internal structure of all the other terrestrial planets described earlier. The thermal balance of planets of this type (including the Moon) depends on their volume to surface ratios which, for spherical globes, are proportional to their dimensions; the larger the planet, the more heat can be contained in its mass. On this basis, the Earth and Venus should contain more internal heat than the smaller planets like Mars, Mercury, or the Moon. We have convincing evidence that the heat engines of the latter, if any, are very weak. Small planets do not possess any measurable magnetic fields (suggesting the absence of inner metallic cores), nor do their surfaces show any mountain chains that could have been produced by folding of their crusts or by other effects of 'plate tectonics' so conspicuous on the Earth and driven by viscoelastic convection in the outer parts of its mantle. Only our Earth seems to possess a crust which exhibits evidence of these processes, with Venus as the next likely candidate.

The Crust of the Earth and its Properties: Hydrosphere and Atmosphere

Now let us emerge from the interior of our planet to its surface and consider in greater detail the layer floating on top of the mantle. 'Floating' is the right word here, for the uppermost 30–40 km of the Earth constitute a shell of 'slag' exuded by the mantle and consisting of rocks which have floated to the top in the course of time because of their lower density. This shell, whose thickness is not equal all over the globe, represents the 'crust' of our planet. The aim of this section will be to outline some of its salient features.

By mass, the crust represents less than 0·01% of the Earth as a whole, and consists of rocks whose average density ranges from $2·8$ g cm^{-3} in protruding crustal blocks (the 'continents') to $3·3$ g cm^{-3} in layers underlying these continents or ocean floors. Laboratory analysis has established that the principal chemical constituents of these rocks are oxygen (about 47% by weight), silicon (28%), and aluminium (8·4%), followed by iron (2·5%), calcium (2·4%), and other elements in diminishing amounts. As far as mineralogical composition is concerned, the continental areas consist predominantly of granites (quartz, orthoclase, mica, etc), while the underlying layers are basaltic (gabbro, olivine, plagioclase, etc) in character. The dividing zone between the bottom of this basaltic layer and the top of the underlying mantle (the 'asthenosphere') is marked by a further increase in density, usually referred to as the 'Moho discontinuity' in honour of Andrija

Mohorovičić, a Yugoslav geophysicist who first inferred its existence from seismic evidence.

Unlike the 'asthenosphere', which is no longer strictly at rest but capable of slow and persistent motions powered by internal heat, the crust floats passively on the underlying denser substrate in accordance with the requirements of hydrostatic equilibrium (or, as geophysicists prefer to term it, the 'principle of isostasy'). How is it possible, we may ask, for one stone to float on top of another merely because its density is less than that of its substrate? Does not seismic evidence disclose that the Earth's crust transmits both the 'pressure' and 'shear' waves excited by earthquakes, a fact which indicates that it behaves like an elastic solid rather than a liquid in which objects could float? Strange as it may seem at first sight, the mantle of the Earth sometimes behaves as if it were solid and liquid at the same time, depending on the duration of the stress to which it may be exposed.

To give an example of such behaviour under conditions which are familiar in everyday life, suppose that we take a stick of ordinary sealing wax and hit it with a hammer. The stick will, of course, break into many pieces as if it were a piece of glass. But if we put another stick of the same wax in a jar and forget about it, in a year or so we shall find that it has spread all over the bottom of the jar as if it were a liquid. In the same way, a metal coin placed on a seemingly solid surface of tar will sink through it if given enough time, while a piece of cork will move upwards through the 'solid' tar and eventually float upon it as if it were water. All solid materials which are not crystalline can react to external stimuli as if they were rigid or fluid, depending only on whether the duration of the disturbance is short or long in comparison with what physicists call the 'Maxwellian relaxation time' of the respective substance. For the rocks found in the Earth's interior the 'Maxwellian relaxation time' amounts only to a few days, or a week at most. When the terrestrial crust and the mantle receive an impulse from an earthquake lasting only minutes, they will respond by vibrating like an elastic solid. But to a permanent force—such as the centrifugal force produced by diurnal rotation—the Earth will respond by becoming flattened at the poles to exactly the same extent as if its entire mass were fluid.

For similar reasons, the asthenosphere itself is disturbed from the state of rest by slow currents of viscoelastic convection driven by the heat engine deep inside the interior of the Earth. These motions are generally very slow—of the order of 0·1–1 mm per annum—and are bound to be shared by any crustal material that 'takes a ride' on top of the asthenosphere. In its exposed position, and being more rigid, the asthenosphere could not retain the consistency of an unbroken shell and was split up into a number of blocks or *plates* capable of independent action. In fact, *it is the relative motions of these plates—propelled by the underlying asthenosphere—that are responsible for the main features of terrestrial geography*. The continental land masses, which are thousands of kilometres wide and about 20–30 km thick, float slowly on the denser substrate like a pancake carried by a slow stream, but may not always be able to do so everywhere with the same speed. A non-uniform translation may cause them to warp along lines where they encounter any resistance, such as they surely will if they happen to be on a collision course with another continental plate.

The effects of these encounters, albeit gradual, may entail dramatic consequences: a part of one continent may be submerged by the other and thus be 'consumed' by the underlying layers; or their collisions may give rise to chains of fold mountains so characteristic of the terrestrial landscape but so conspicuously absent on the other planets described earlier. A typical example is the semi-continuous chain which begins in Europe in the Bay of Biscay with the Pyrenees, runs through the Alps and Carpathians, and continues through the Caucasus to Pamir and the Himalayas in Asia. Another example is the great chain of the Rocky Mountains and the Andes which runs from Alaska to Patagonia along the entire west coast of the American continent. Perhaps the most dramatic recent tectonic encounter between continental plates is represented by the Indian subcontinent running from the south into the main Asiatic land mass (colour plate 9). This collision has raised the Himalayas to their exalted heights and the after-effects continue to act at the present time (the elevations of the Himalayan giants are still on the increase). Similarly, regions of tectonic instability manifested by volcanism and earthquakes, such as those girdling the shores of the Pacific basin from Japan through Alaska to South America, are now also known to be connected with encounters between continental plates and arise from the accompanying mechanical effects.

How old are the rocks constituting these and other formations of the Earth's crust? Radiometric determinations (cf pp 84–6) of their ages have disclosed some interesting facts, one of which is that *most rocks found on the Earth are approximately 10 times younger than those on the Moon* (Chapter 3). Since both these celestial bodies must be very nearly the same age, the only reasonable explanation for this difference would seem to be a conclusion that *most rocks now forming the crust of our planet were not part of its original surface, but emerged from the interior to solidify at a later date.* Partial exceptions to this rule are known (cf the next section), but by and large the radiometric ages of both the continental areas and the ocean floors do not appear to exceed a few hundred million years. This fact constitutes an eloquent testimony to the speed with which the mantle of the Earth consumes its crust and remelts rocks at elevated temperatures before disgorging them back to the surface where they solidify again.

Up to now we have been describing the crust of the Earth as if it were all solid land mass. This, of course, is not quite true, for about two-thirds of it are ocean basins filled with *water*, a property which renders our Earth unique in the solar system. The basins themselves represent shallow, yet extensive pockmarks on the face of the Earth; their maximum depth is close to 11 km, which exceeds the altitude of the highest mountain on land (Mount Everest, 8884 m high) by only 2 km. Just as the highest mountains occur largely in chains (like the Himalayas with their string of peaks over 8000 m high), the greatest abyssal depths of the oceans are found in relatively narrow trenches (like the Marianna or Tonga Trenches in the Pacific, or the Puerto Rico Trench in the Atlantic). In fact, 11 kilometres represent less than a six-hundredth of the Earth's radius and less than one-half of the amount of polar flattening produced by the daily rotation of our planet. If pockmarks as shallow as the ocean beds were marked to scale on a globe one metre in diameter, both the Atlantic and the Pacific, not to mention the smaller oceans, would be represented by depressions barely one millimetre deep.

When examined in greater detail, however, the ocean floors are far from smooth. If we could siphon off all the water from the Atlantic, we would find on its floor the longest mountain chain on the Earth—the famous mid-Atlantic range running in a north–south direction with a deep central crack. The origin of the Atlantic range seems to be volcanic, and is probably connected with continental drift; no other range on the Earth can compare with it in length or average height. The highest mountains on the Earth are, incidentally, not to be found in the Himalayas, but actually rise from the floor of the Pacific. The famous Hawaiian volcanoes Mauna Kea and Mauna Loa in the mid-Pacific attain altitudes of only 4208 and 4170 m above sea level, respectively; but since they rise from an ocean floor 5800 m below sea level, their actual altitudes above the surrounding abyssal plain are close to 10 000 m, in other words, more than 1100 m higher than the altitude of Mount Everest above sea level. Or, to give another example, the Pico Islands in the Azores rise 8100 m from the floor of the Atlantic, though only their topmost 2300 m emerge above sea level.

The ocean basins on the Earth are, of course, filled with water, the ensemble of which constitutes the terrestrial *hydrosphere*. The total mass of this water is only a little more than a ten-thousandth of the mass of the Earth as a whole, and if it were distributed uniformly all over the globe, it would cover the Earth with an ocean about 1800 m deep. Chemically, ocean water is far from pure; each cubic kilometre (weighing 1 billion tons) contains, on average, 19 million tons of dissolved chlorine, 10·6 million tons of sodium, 1·3 million tons of magnesium, and proportionally smaller amounts of other elements† which make it truly 'salt' water, both chemically and by taste.

Where did all this water come from? We are fairly sure that the new-born Earth contained no water on its surface—it was born 'dry'—and that its oceans were later exuded from the interior by the thermal cracking of hydrates. Let us try to explain what we mean. The molecule of water (H_2O) is a cosmically common and very stable product of nature. We have found by means of its radio-emission that it exists in interstellar space in the form of tiny ice crystals, and we know of many minerals on the Earth—such as obsidian and many other kinds of volcanic glasses—which contain as much

† Among them we would also find about 300 kg of silver and 4 kg of gold!

as 10% of water molecules by weight in their structure. Mineralogists call such specimens 'hydrates'.

At the normal temperatures prevalent on the surface of the Earth the hydrates are generally stable and can retain their water for an indefinite period of time. Moderate heating to temperatures between 500 and 1000 °C is, however, sufficient to break the chemical bonds holding the H_2O molecules within the molecular structure of the hydrates and expel them in the form of steam which can condense into water. Temperatures of this order are likely to have been exceeded in most parts of the Earth's interior. Thus the former hydrates contained there are already largely despoiled of their water supply. If water was originally present in the primordial material from which the Earth was formed in solid state to an extent of 1 part in 10 000 or more, the thermal cracking of the hydrates, and outward seepage of water as superheated steam, could have eventually produced as much of it as we find in all the oceans of the Earth.

How old are the oceans? While nothing is known about the possible hydrosphere of our planet in the first 'dark aeon' of its existence, the oldest preserved rock strata—more than three billion years of age—already bear evidence of the existence of 'sediments' which required water for their formation. Moreover, the original 'juvenile' water squeezed out by heat from the Earth's interior probably contained a few dissolved salts before it emerged to the surface. Chemists estimate that, in order to acquire its present degree of salinity, sea water must have been leaching its bedrocks for about three billion years—a fact important for the emergence of life on our planet (on which more will be said later).

In reference to the terrestrial hydrosphere, we should not forget to add that an appreciable (though not large) part of the terrestrial water is confined to solid state in the *polar caps*. In fact, as approximately 70% of the terrestrial surface is covered by both sea and fresh water, about 7% of that surface is permanently covered with *ice*. Its main depository is the Antarctic ice sheet, which is several kilometres thick and covers an area in excess of 12 million km^2 around the south pole. Together with the surrounding pack-ice, the total ice-covered area may reach 18–20 million km^2 in wintertime. The Arctic ice cap contains somewhat more than one-half of the southern supply of ice; and if both were to melt completely, the global level of the oceans would rise by more than 160 m. In this respect, the Earth again stands apart from her planetary sisters, for while the polar caps on Mars also contain a certain fraction of water, its amount in frozen state forms a layer only centimetres deep (see p 119). Far from surviving the yearly cycle, the material of the Martian polar caps sublimes, rather than melts, into the environment with the advent of spring. In contrast, the terrestrial polar caps constitute secularly permanent features of our planet; and with the high reflectivity of their material, they would be very conspicuous landmarks for extraterrestrial observers.

The continents and hydrosphere (both solid and liquid) together form the topmost layer of our planet and constitute an outer skin, so to speak, so thin that its depth does not exceed 0·2% of the radius of the Earth. Surrounding this layer is a gaseous envelope known as the *atmosphere*, whose presence above the surface is also not accidental: in fact, it represents yet another aspect of the same process which endowed the Earth with its hydrosphere. If the hydrosphere originated by de-fluidization of the Earth caused by a gradual build-up of radiogenic heat in its interior, the atmosphere originated by its degassing—that is, through the liberation from the interior of volatile elements whose molecular weights were large enough to prevent their escape from the Earth's gravitational field. The gradual formation of the hydrosphere and atmosphere represents, therefore, two complementary aspects of the same process, for an atmosphere capable of exerting adequate pressure is necessary for the maintenance of any liquid on the surface. Planets that possess atmospheres but no hydrospheres can exist —for example Mars and Venus—but the converse is physically impossible.

The chemical composition of our atmosphere is well known, and is in agreement with the ability of the Earth's gravitational field to retain gases of given molecular weights. The principal constituents are nitrogen (75·5% by weight) and oxygen (23·1%), followed by argon (1·3%), neon (13 ppm), helium (0·7 ppm), and diminishing amounts of heavier inert gases, such as krypton and xenon, present in quantities so minute that a chemist would describe them as impurities. Variable amounts of water vapour (0·01–0·1%) and carbon dioxide (0·03%) make up the rest. Hydrogen, a principal constituent of water in the hydrosphere, is too light a gas to be permanently retained in the atmosphere and occurs in its outer layers in barely significant amounts. Apart from water vapour and carbon dioxide, the composition of the atmosphere is remarkably uniform up to a height of at least 100 km—a fact which testifies to the efficiency with which atmospheric gases with different molecular weights are intermixed by atmospheric turbulence.

The total mass of the terrestrial atmosphere is $5·3 \times 10^{18}$ kg. It constitutes, therefore, only about one-millionth of the mass of the Earth as a whole and less than 0·3% of the hydrosphere. An atmospheric column of 1 cm^2 cross section at sea level weighs about 1·034 kg and exerts an average pressure capable of supporting a column of mercury about 76 cm in height. The volume of space occupied by the atmosphere is, however, more impressive, extending as it does to an altitude of several hundred kilometres. The bulk of the atmosphere is confined (by self-compression) to its base and is described by meteorologists as the *troposphere*, which extends to a mean altitude of 11–14 km depending on the latitude. The highest mountains on the Earth just approach the top of this layer, and modern jet aircraft skirt it in cruising flight. Above it lies the *stratosphere*, so called because the decrease in temperature throughout the troposphere is arrested (at some 218 K); at higher, stratospheric levels the temperature actually begins to increase.

At altitudes of 60–80 km we encounter another distinct layer—the *ionosphere*. At these heights the ambient air pressure diminishes to only about one part in 5×10^5 of that at sea level, and the air density is only about $9 \times 10^8\,\mathrm{g\,cm^{-3}}$—a value which at sea level is close to $0\cdot0012\,\mathrm{g\,cm^{-3}}$. Air rarefied in this way is easily ionized by energetic radiation from the Sun, and the number of electrons at these heights may amount to as many as 10^6 per cm^3. The ionosphere is composed of several layers and extends almost up to 300 km, where it acts like a metallic mirror or mesh that can reflect high-frequency radio waves downwards; it can also backscatter similar waves reaching our Earth from space.

The height of the ionosphere, as well as its transparency, oscillates diurnally, and its outer surface may attain altitudes of nearly 300 km. The temperatures prevalent at these levels vary and may attain values far in excess of those encountered on the surface. The density of such a hot gas ranges, however, only between 10^{-11} and $10^{-14}\,\mathrm{g\,cm^{-3}}$. These regions can give rise to beautiful displays of polar aurorae, and even in their lower layers the density is sufficient to burn up the majority of meteors by atmospheric resistance (see Chapter 6).

Above the ionosphere, the remaining air is so rarefied, and the mean free path of the individual particles so long, that we are dealing, no longer with gas, but with an assembly of individual particles that describe ballistic trajectories of increasing length between individual collisions which become progressively less frequent. This is the terrestrial *exosphere*, through which our planet borders on interplanetary space; and some of its more energetic particles are constantly being lost to space whenever their speeds exceed the velocity of escape from the Earth's gravitational field.

In another sense the astronomical domain of the Earth does not end at the neutral exosphere. Far above it we encounter condensations of charged particles (mainly protons and electrons) trapped by the magnetic field of the Earth and known as the terrestrial van Allen belts in honour of their discoverer, or, more generally, as the terrestrial *magnetosphere*. This forms a doughnut-shaped ring around the Earth (see figure 11) and is populated by charged particles of partly solar, and partly interstellar, origin.

Theoretical studies of the motion of charged particles in the vicinity of the terrestrial magnetic dipole were initiated many years ago by C Störmer, who pointed out the discrete nature of the different belts that the charged particles of external origin may, or may not, enter. Possibly the first suggestion that the accessible Störmer regions may actually be populated came from H Alfvén in 1947. The existence of corpuscular radiation trapped in their inner regions had already been indicated by experiments aboard the Russian Sputnik 2 in November 1957. However, confirmation came from the American Explorer 1, which was the first man-made space probe to actually penetrate the interior of the first (close) radiation belt in the early part of 1958; and a correct interpretation of its data was advanced by van Allen in May of that year.

The geometry of the van Allen belts is illustrated in figure 11. The inner ring, shown in cross section as two lobes, extends radially from about 1000 to 10 000 km above the Earth's surface and contains a relatively large proportion of high-energy protons. The outer ring is much more voluminous—extending as it does from some 25 000 to 60 000 km in the radial direction—and much less densely populated. It contains mainly electrons moving about more slowly, but still possessing energies far exceeding the thermal energies of particles in the exosphere. Both the protons and electrons orbiting (or librating) in the terrestrial dipole field originate largely from the solar wind, though the protons and electrons in the inner belt may also be produced by interactions between molecules in the extension of the terrestrial exosphere and cosmic rays.

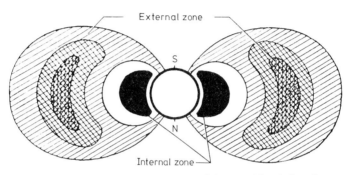

Figure 11. A schematic representation of the van Allen belts of geomagnetically trapped charged particles around the Earth.

The particle flux inside both belts fluctuates with solar activity, being of the order of 10^3 particles per cm^2 per steradian in the outer belt and exceeding 10^6 particles in the inner one. The geometrical cross sections of both belts, and especially the outer one, vary considerably in the course of time. Their regions are, in fact, not sharply defined, for trapped corpuscular radiation must be spread throughout the Earth's magnetic field. The inner and outer van Allen belts should, therefore, be regarded merely as regions where the space density of moving charged particles is greater than elsewhere.

This completes the brief survey of the anatomy of our planet. We have found the Earth to consist of a (probably) solid inner core of nickel–iron. The core has a radius of 1250 km, is surrounded to 3470 km by a liquid shell consisting of much the same material (with possibly some admixture of molten silicates), is further overlain by a mantle of silicate composition, and is covered by a crust. This crust supports the hydrosphere, and is protected by a gaseous atmosphere of several distinct layers gradually petering out through an exosphere into a magnetosphere, which represents the physical boundary of our planet. Although the solid globe has a mean radius of only 6371 km, the limits of our magnetosphere may extend at times almost 10 times as far from the Earth's centre in the direction of its magnetic equator. Only when we reach this point do we complete the physical portrait of our mother planet.

The second characteristic feature of the Earth to be stressed is the fact that its model outlined above is far from static on a cosmic timescale; so far its activity has shown no signs of calming down. This is certainly true of the Earth's interior whose core is probably still on the increase and whose incessant activity is borne out by the vagaries of the magnetic field emanating from it. But it is, in particular, true of the crust and its protective layers of air and water which have never stopped moving since they were formed. The motions of the crust due to plate tectonics, which continuously reshape the terrestrial geography, are powered by the internal heat of the Earth, and appear sedate to us only because we view them on too short a timescale. But the perpetual motions in the hydrosphere and atmosphere, powered by the Sun, are changing the face of the Earth from day to day and from season to season, thus making our planet the veritable beauty-queen of the solar system! What a contrast to the petrified faces of the Moon or Mercury, grey with age, on which nothing ever happens for astronomically long periods of time. And is not the Earth's exquisite beauty, as seen from afar (colour plate 8) as well as at close range (colour plate 9), a sufficient reason to cherish our planet and protect it from any wanton harm?

History of the Earth and its Future

In the preceding sections of this chapter we sketched a profile of the interior of our planet and, in particular, of the structure of the outer parts of its mantle. The aim of the present section will be to place its evolution in cosmic historical perspective and, by extrapolation, anticipate its probable future.

The origin of the Earth, as of other planets, represented an event which can be dated with considerable accuracy, but whose circumstances are still known only vaguely. That our planet came into being as a part of the same creative act which produced the Sun and the rest of its system is undoubted. The cosmic date of this act can be inferred from the radiometric ages of chondritic meteorites, as well as of the lunar soil, to have taken place 4·67 billion years ago, with an uncertainty affecting only the last decimal of this number by one or two units. Again from radiometric evidence, we know that the actual duration of this act was no longer than a few million years—a time fleetingly short in comparison with the span of the subsequent history of the system.

But when it comes to the circumstances under which the Earth and the other planets were born, we are still very much in the dark. We do not know with any assurance the mass of the primordial 'solar nebula', nor the temperature at which it condensed into the planets and lesser bodies described earlier in this book. We do not know the number of planets and other denizens of our solar system that existed at that time. We know only that planetary globes must have originated by a condensation of material in *solid* state, for their masses are too small for the retention of gas by gravitational attraction. But at least this fact imposes an upper limit to the temperature of the medium from which the planets originated.

As far as the Earth is concerned, the first billion years represent the veritable 'dark aeon' of its existence, for no direct evidence of its state at that time has been preserved in the strata of its crust. The oldest rocks of the crust are some 3·6–3·7 billion years old and have been preserved only in rare localities, for example and in particular, in northern Greenland. At the time when these rocks solidified, a major part of the active history of the surface of the Moon was already over (Chapter 3), but on the Earth it had scarcely begun.

We cannot say much about the history of the terrestrial interior and its differentiation up to that time. If the Earth accumulated as a solid body of approximately homogeneous chemical composition, a considerable time must have elapsed before a gradual build-up of radiogenic heat in the interior progressed far enough to begin extraction of iron (and other metals) from the rocks. During this time drops of metals were pulled down by gravity to form eventually the present core of our planet; and even as this was going on, the lighter 'slag' floated to the top of the core to form the primordial mantle and the crust. It is, of course, possible that a differentiation of the Earth into its core and mantle had already begun in the pre-solar past—within the 'solar nebula'—with condensations of metals providing accretion centres for subsequent condensation of silicate material. But this is unlikely, and the principal reason why is provided by the existence and composition of the terrestrial hydrosphere and atmosphere. In order to explain this, perhaps a few words of description would be of use here.

Those parts of the terrestrial surface that have not been 'consumed' by the continental land masses have rocks (in Greenland, the Canadian shield, or the South African plateau) 3·4–3·7 billion years old. Although these rocks are greatly metamorphosed, they are *sedimentary*, a fact from which it transpires that *water in liquid form must have already existed on the terrestrial surface at that time*. In turn, the existence of water implies the presence of an *atmosphere*, whose pressure protected the liquid and prevented its dispersal into space.

The full significance of these facts will emerge if we recall that, 3·5 billion years ago, the youthful Sun was not bright enough to warm the terrestrial surface above the freezing point of water if the composition of its atmosphere had been anything near to what it is today. In order to exert a sufficient 'greenhouse effect', the atmosphere would have had to consist mainly of the compounds of hydrogen—methane, ammonia, and water vapour—and contain large amounts of carbon dioxide as well. The primordial oceans, at that time shallow and lacking the salinity which they subsequently acquired by the gradual leaching of their bedrocks, probably covered a larger proportion of the terrestrial surface than they do today. The microscopic structure of surface rocks, however, exhibits indications which palaeontologists regard as fossil remnants of *living* organisms. The so-called 'stromatolites', which are reef-like remnants of the precipitation of microbial communities, date back 3·4 billion years and may constitute the first fossil imprints of the earliest life on our planet. Certain other fossil-like microstructures found in Swaziland in Southern Africa may again represent our planet's first flora.

For a long time the subsequent evolution of these primitive forms of life was very slow. But about two billion years ago an event took place which profoundly changed both the future course of life and the entire terrestrial environment: namely, the apparently sudden emergence of *green plants*. During the preceding *Archaean Age* which extended to more than half of the present age of our planet, nature had obviously not yet made up its mind as to the type of living matter likely to prove most successful under terrestrial conditions. At least bacteria still exist which thrive on the decomposition of methane and to which oxygen is, in fact, lethal. These bacteria are possibly direct descendants of the oldest inhabitants of our planet and have an ancestry that goes back to the Archaean Age.

Some time around two billion years ago, however, nature must have found out by a long, drawn-out process of trial and error that the radiant energy from the Sun could be utilized most effectively by the *photosynthesis of carbon dioxide* (of which there was an ample supply in the Archaean atmosphere) into other, more nutritious compounds by means of a magic substance called *chlorophyll*, a pigment responsible for the green colour of plants. One of the most important by-products of this process was the release of large amounts of molecular oxygen into the atmosphere.

The first protagonists of this planetary revolution were green *algae* and they still flourish in our waters today. It is mainly through their action that the atmospheric reserves of primordial carbon dioxide were rapidly consumed and replaced by molecular oxygen. This, in turn, altered both the optical properties of our atmosphere (by preventing all ultraviolet radiation from the Universe from reaching the ground) and the chemistry of the continental rocks formed since that time. Archaean rocks, like those on the Moon, still contain oxides in the lowest state of oxidation, while rocks formed later are more highly oxidized.

The apparently sudden change in chemical composition of our atmosphere (i.e. from a reducing to an oxidizing one) was caused by a rapid proliferation of green algae and marked the beginning of the *Proterozoic Age* in the long history of our planet. In the course of this era organic life began to make its presence felt more and more insistently in the terrestrial environment. The 'feeding on air' made possible by the practice of photosynthesis opened up an entirely new possibility for life's subsequent evolution on the Earth. About 1·5 billion years ago, it led to the development of nucleated cells of plant life, and somewhat later—about a billion years ago—to that of sexual differentiation. Moreover, in the last 600–700 million years nature acquired, and eventually mastered, the art of formation of *polycellular organisms* and the way to make an agglomeration of cells work in unison. This led to the emergence of macroscopic life—Metazoa—the different species of which were capable of further specialization; these have since come to dominate the biological scene of our planet.

Biologists, and many others, have never ceased to wonder at the speed with which the Metazoa seem to have come into being at the opening of the *Phanerozoic Age*, a period in the history of our planet directly before the present. Its opening act—the Palaeozoic Era—began 600 million years ago when the layers of Cambrian rocks were being deposited. Their testimony discloses that living organisms as complex as various types of brachiopods and trilobites already dominated the shallow waters of Cambrian (and Ordovician) times. However, with the opening of the Silurian period some 440 million years ago, invertebrate organisms began to lose their leading position on the evolutionary ladder: the first fish appeared, and green plants began to gain footholds on dry land. True seed plants appeared in the subsequent Devonian period 350–400 million years ago. It was in this period also that animals left the maternal womb of the shallow waters to take their first tentative steps (as Arachnida and Insecta) towards proliferation on dry ground. The exodus did not, in fact, leave the waters bereft of life. Bony fish, including air-breathing forms, remained abundant in fresh waters, and amphibians began to make their appearance towards the end of that period.

At the next wave of the fairy wand the scenery changed again. During the Carboniferous period, 280–350 million years ago, the landscape became covered with swamps dominated by giant ferns and vegetation which offered ample homes not only to the abundant insects, but also to the amphibians and the first reptiles. Petrified forests of luscious vegetation produced, in time, extensive seams of anthracite and hard coal which have given this period its name. The following Permian period, 230–280 million years ago, closed the Palaeozoic Era. During the Permian, the rapid desiccation of the climate, followed by an 'ice age', brought to an end the long line of trilobites on the Earth and greatly reduced the life in the shallow waters in species and in numbers.

On the other hand, the opening of the Mesozoic Era in the Triassic period, some 220 million years ago, once more brought about an efflorescence of life on this planet and ushered in the age of the giant reptiles. The subsequent Jurassic and Cretaceous periods saw the development of dinosaurs, the largest living creatures that ever walked on the surface of the Earth. They populated the tropical forests and fed on vegetation that was rapidly approaching the form of our contemporary plants and trees. Plane trees, sycamores, oak, walnut, and palm trees offered food and nesting places to the first birds, and the first small mammals made their furtive appearance only in the dead of night.

A change in global temperature towards the end of the Cretaceous period heralded the advent of a cooler climate which saw the extinction of the dinosaurs and other large reptiles. The opening of the Tertiary Era in the Palaeocene some 60–70 million years ago introduced on this planet the age of the mammals and of vegetation differing only a little from that which we know today. The climate became warmer in the Oligocene (26–38 million years ago), but grew temperate again in the Miocene (5–25 million years ago). By that time, the terrestrial continents had more or less assumed their present shape, and the grasslands had increased markedly in the world at the expense of the forests. And the tropical forests of the Old World abounded with many kinds of monkeys and apes, some of which became the distant ancestors of the human race that dominates our planet today.

The roots of man's success story in the Miocene period are indistinct and based on very fragmentary information, though anthropologists now claim that no longer is any essential link in the whole evolutionary tree of man missing. The story assumes a more coherent form during the last few million years of the Pliocene period; and its continuation into the Quaternary Era is too well known to need any repetition in this place.

In the course of these ages, dry land was distributed on the terrestrial globe very differently from what it is today. For a long time it consisted of only one continent ('Pangaea'), floating on top of a denser basaltic substrate by virtue of isostasy, and propelled to move by slow convection currents deeper down in the mantle. The disintegration of Pangaea into the separate continents and subcontinents of the present-day geography occurred only in the relatively recent past. North America was separated from the present Europe only in the Triassic period at the beginning of the Mesozoic Era some 220 million years ago. South America became separated from Africa somewhat later (186 million years ago), the isthmus connecting it to North America being created much later by volcanic activity. The Mediterranean Sea—that favourite haunt of North-Europeans in search of the sun—represents a late-Tertiary creation only about 25 million years old; the 'last touches' to it were added during the last ice age and its aftermath in the last 10–20 000 years.

The geological history of the Earth contains in its stony strata evidence of other transient events which have left their mark on the terrestrial landscape. Conspicuous among these are successive periods of mountain formation, three of which are on record in the Phanerozoic Age: the Caledonian and Hercynian folding (so called after the principal localities of their remnants) which occurred in the Lower Devonian and Permian periods of the Palaeozoic some 400 and 250 million years ago; and lastly, the Alpine folding which started in the mid-Tertiary (only 25–30 million years ago) and is not yet over even at the present time. All of them have been connected with the relative motions of continental land masses. As a result of their collisions, the Alps have risen by 2 km and the Himalayas by 3 km in the past

10 million years. The Sierra Nevada on the west coast of the United States has risen by 2 km in only 2 million years! The enormity of the forces engaged in mountain formation can be illustrated by the fact that rocks containing fossils of creatures which lived in the sea have been lifted in a few million years almost to the altitude of Mount Everest!

Other, almost-periodic features of the terrestrial book of hours are recurrent *ice ages*, in the course of which the continental climate became noticeably cooler for several million years. The Quaternary ice age, which descended on the northern hemisphere in the past few million years (and an interglacial period in which we happen to find ourselves at the present time), is too well known to require much deliberation here. It may, however, be less well known that it was preceded by another ice age in the Permian period some 220 million years ago, and by yet another one in the Silurian period over 400 million years ago. Ice ages are also known to have occurred on our planet before the beginning of the Phanerozoic Age. Particularly good evidence for one towards the end of the Archaean Age (some 2300 million years ago) has been preserved in the old strata of the Canadian shield.

The real cause of these recurrent ice ages is not yet known with any assurance. The scarcity of evidence for the earlier ones does not rule out that they were likewise produced by continental drift when, in the course of time, the continent affected wandered too close to the poles of our planet. Certainly the regions of Antarctica now must once have known much milder climates for carboniferous vegetation to have provided sufficient seams of coal at such high latitudes.

Other causes may have cooperated to produce the observed effects of glaciation. But one fact is certain: during each ice age large amounts of water were withdrawn from the oceans and were used to cover both the polar regions and large continental areas at more moderate latitudes with ice sheets. Thus, during the early (warm) Tertiary Era some 30 million years ago, the global ocean appears to have been about 70 metres above its present level, while at the peak of the last (Quaternary) ice age it sunk to a level 130 metres below what it is now. Of course, such large changes in sea level altered the outlines of the present continental land masses almost beyond recognition. Suffice it to say that only some 18 000 years ago Great Britain was a part of Europe (whether the local patriots at that time liked it or not), with the North Sea dry land and the River Thames a tributary of the Rhine. Less than 7000 years ago one could have crossed the Straits of Dover on foot! It was only with the advent of a milder climate in the present interglacial respite that Northern Europe and Scandinavia were relieved of their ice sheets, and in response to the requirements of isostasy these lands have been rising ever since.

And this takes us to the face of the Earth as we know it today. What will it look like in the future? There is no doubt that forces which operate in the Earth's crust and act on its exposed surface will continue to do so and cause further ice ages to come and go. The continuing continental drift will give no rest to dry land and will keep disrupting it into subcontinents and islands of progressively smaller size.

Political complications which may arise in the course of these processes are wholly outside the scope of this book. As far as physical processes are concerned, the gradual warming up of our planet by radiogenic heat released by the spontaneous disintegration of long-lived radioactive elements is, however, likely to continue for at least the next two to three billion years. Not much juvenile water from the interior will enrich our oceans by future desiccation of the terrestrial mantle, nor can our atmosphere expect much replenishment by its degassing. This means, therefore, that in the future our climate will be increasingly at the mercy of heat provided by sunlight—a fact which spells nothing good for our descendants. We have already mentioned that, in the past, the Sun was appreciably less luminous than it is at the present time. Astrophysicists studying the evolution of the stars (prompted by irreversible nuclear changes taking place in stellar interiors) are convinced that, with diminishing reserves of hydrogen in its deep interior, the Sun will gradually become hotter in the future. This trend must eventually make the surface of the Earth so hot that its oceans will evaporate and its atmosphere will disperse into interplanetary space, thus exposing the naked surface of our planet to a scorching heat which no form of life could possibly survive.

This is especially true because, by that time, a gradual slowing down of the axial rotation of the Earth by lunar tides (cf p 95) will have prolonged our day to approximately two months of our present time, so that there will be no more than six days in a year. This implies that the climatic extremes on the Earth may eventually approach, or even exceed, those prevalent on Mercury today, and no-one would ever consider looking for any vestige of life on that planet!

Thus the cosmic episode which began some 4·6 billion years ago with the formation of our solar system is destined to meet its end in the scorching breath of a merciless Sun smarting under an increasing scarcity of hydrogen in its deep interior. There is nothing that we, or the other planets, can do that will rescue us from the Sun's ultimate fiery embrace. The only comforting aspect of this pessimistic outlook is the fact that the dismal end is still a long time in the future. Very probably another five to seven billion years will elapse before sunshine changes from a benign friend to a relentless enemy. Events may be either accelerated if, in the course of time, the planets move closer to the Sun to meet their doom in its glare, or delayed should the planets slowly spiral outwards under the command of the inexorable laws of celestial mechanics.

The story of our solar system runs from the oldest rocks on the Moon—which probably saw the Sun shrink towards its Main Sequence stage—and from the time when the first manifestations of life flickered in the shallow waters of the Earth, to the most exalted heights which its eventual progeny—temporarily represented by human beings—may attain in the future. By the time the end comes, this story will have lasted for more than 10 billion years. Surely this should be long enough to endow the particular cosmic episode described in this book with some significance in the annals of our Galaxy!

Index

Adams, W S, 181
Aldebaran, 37
Aldrin, E E, 69
Alexander the Great, 43
Alfvén, H, 14, 208
Allen, C W, 170
Ammizaduga, King of Babylonia, 173f
Anders, W A, 3
Andromeda Nebula, 163
Antarctic, 2
Aphrodite, 173, 195
Archimedes, 8, 44
Aristarchos of Samos, 7f, 40
Armstrong, Neil A, 3f, 69
Artemis, 39, 173
Asteroids, 10, 80, 151ff
    albedo of, 156
    axial rotation of, 157ff
    commensurabilities of, 154
    Hornstein–Kirkwood gaps, 154
    mass of, 157
    orbits of, 153f
    shape of, 157f
    Adonis, 62, 154f, 157
    Apollo, 155, 157
    Astraea, 153
    Ceres, 152, 154ff, 157f
    Chiron, 155
    Eros, 62, 155, 157f
    Eumeia, 156
    Geographos, 155
    Hebe, 156
    Hermes, 62, 155, 157
    Hirayama families, 154
    Hygeia, 156
    Icarus, 154, 157
    Iris, 156
    Juno, 153, 156
    Metis, 156
    Pallas, 152, 154ff
    Psyche, 156
    Themis group, 154
    Trojan, 154, 156
    Vesta, 153, 155ff
Astronomical unit, 7, 177

Baltimore Gun Club, 39
Barbicane, Impey, 39
Beethoven, L van, 39
Benedict, W S, 186
Blackwell, D E, 168
Borman, F, 3
Brahe, Tycho, 9, 41
Braun, W von, 22

Cadiz, 1
Camichel, H, 175
Cape Kennedy, 22
Carter, J L, 75
Cathay, Islands of, 1
Columbus, Christopher, 1f, 43
Comets, 10f, 80, 141, 166
    Biela, 166
    Giacobini–Zinner, 166
    Tempel, 166
    Tuttle–Swift, 166
Connes, J, 186
Connes, P, 186
Copernicus, Nicolaus, 9, 99
Curie point, 108

Dareios (Great King), 43
Dunham, Th, 181

Earth, 1f, 4, 7ff, 23, 27, 30, 41f, 44ff, 51f, 59f, 64ff, 71f, 74, 87, 91f, 94ff, 99, 102ff, 110ff, 117, 129f, 136, 139f, 157, 161, 164, 166f, 170, 172f, 176, 178ff, 182, 187, 189, 191f, 195ff
    Archaean Age, 212, 215
    asthenosphere, 202f
    atmosphere, 207f
    Cambrian period, 96, 213
    Carboniferous period, 96, 213
    chronology of, 211ff
    Cretaceous period, 78, 96, 213
    crust of, 59, 192

Devonian period, 95f, 213
dimensions of, 196
hydrosphere, 204f
internal structure of, 197ff
internal temperature, 200f
ionosphere, 208f
Jurassic period, 213
life on, 212ff
mass of, 196f
Mauna Kea, 139, 205
Mauna Loa, 139, 205
Mesozoic Era, 96, 213f
Miocene period, 214
Oligocene period, 214
orbit of, 195
Ordovician period, 213
origin of, 210f
Palaeocene period, 214
Palaeozoic Era, 83, 94, 96, 213
Pangaea (continent), 214
Permian period, 213, 215
Phanerozoic Age, 213ff
Pliocene period, 214
Proterozoic Age, 212
Quaternary Era, 96, 214f
rotation of, 195f
Silurian period, 213
Tertiary Era, 214f
tidal friction, 95, 196, 216
Triassic period, 213f
van Allen belts, 208f
Earth–Moon system, 43, 47, 63, 94ff, 161
Earthquakes, 199, 203
Ebert's rule, 55, 91
Emerson, R W, 13
Eratosthenes of Cyrene, 196
Euler, L, 40

Fastie, W A, 77
Fontana, F, 114

Gagarin, Yuri, 3
Galaxy, 11, 37, 78, 92, 161, 172, 217
Galilei, Galileo, 2, 9, 54, 173
Gallus, Sulpicius, 43
Gault, D E, 76
Gauss, C F, 40, 152
Gegenschein, 10, 167
Gehrels, T, 154
Grand Khan, 1

Hall, A, 143
Hammurabi, King of Babylonia, 174
Harding, K L, 153
Hephaistos, 195
Herschel, William, 3
Hesperus, 173
Hipparchos, 40ff, 44
Homo sapiens, 3
Hooke, Robert, 57, 60, 62
Hornstein, K, 154, 159
Hulst, H C van de, 170

Ishtar, 39
Isis, 39

Jupiter, 4, 8, 10ff, 36, 112, 151ff, 160f, 166, 168f
mass of, 10

Kennedy, J F, 22
Kepler, Johannes, 2, 9, 152
Kepler's laws of planetary motion, 9, 45
Keplerian angular velocity, 171
Kirkwood, D, 154, 159
Klepešta, J, 162
Kopal, Z, 159
Kresák, L, 157, 159

Lagrange, J L, 40, 154
Laplace, P S, 40
Leibnitz, W G, 9
Livius, Titus, 43
Lovell, J A, 3
Lowell, P, 99
Lunar,
    continents, 53f
    craters, 54ff
        Ebert's rule, 55, 90
        'ghost craters', 81f, 89
        microcraters, 74f, 78f
        origin of, 60ff
        ray craters, 82ff, 86
        Alphonsus, 14
        Aristarchus, 82f, 94
        Catharina, 57, 86
        Cauchy, 138
        Clavius, 54f, 62, 81
        Copernicus, 54, 56, 61f, 65, 73, 76, 82f, 94
        Cyrillus, 57, 81, 86, 187

Daguerre, 81
Descartes, 24
Eratosthenes, 82f
Fra Mauro, 24
Giordano Bruno, 82, 85
Kepler, 83
Lambert, 81
Messier, 63
Theophilus, 54, 57, 61, 76, 81f, 86, 94, 187
Tycho, 54, 58, 61, 73, 76, 82, 84, 94
maria, 54ff
  Crisium, 24, 88, 93
  Foecunditatis, 24, 88
  Humorum, 88, 93
  Imbrium, 64f, 76, 81, 88ff, 93, 106
  Nectaris, 81, 88, 90
  Nubium, 57
  Oceanus Procellarum, 24, 82f, 90, 94
  Orientale, 64, 66, 76, 88f, 93, 105f
  Serenitatis, 88, 90
  Tranquillitatis, 24, 63, 88, 90
mountains,
  Alps, 64
  Apennines, 64
  Carpathians, 64
  Mount Hadley, 64
  Taurus–Littrow, 24
rocks, 27, 46f, 51, 58ff, 78, 84
  age of, exposure, 78, 87; radiometric, 84ff, 88
  chemistry of, 17f, 59
  conductivity of, 51
  soil ('fines'), 87
  structure of, 59f, 74
Lunar-roving vehicle, 18, 27
Lyot, B, 184
Lyttleton, R A, 189

McLaughlin, D B, 135
MacQueen, R G, 170
Magellan, Fernandez, 1f
Main Sequence, 160, 193
Mars, 4, 8f, 10ff, 31, 33, 40, 76, 107, 111ff, 151ff, 160f, 164, 166, 169, 177, 182f, 187, 190, 192, 202, 206f
  atmosphere of, 114, 119, 124
  'canals', 125, 131
  canyons on, 137f, 141
  clouds, 114
  composition of, 125
  craters on, 120, 131ff
  dimensions of, 113
  interior of, 139ff
  ionosphere of, 115
  magnetic field, 140
  maps of, 128f
  mass of, 113
  Mons Olympus, 134ff
  orbit of, 111
  period of, 112
  polar caps, 114, 116, 128
  rotation of, 113
  satellites of, artificial, 148; natural (Phobos and Deimos), 93, 143ff, 156f
  shape of, 130
  surface of, 33, 35f, 122ff, 126f, 132
  Syrtis Major, 128
  temperature of, 117
  Tithonius Lake, 135f, 139
  Valley of the Mariners, 135f, 139
  volcanoes, 135
  winds, 121, 124, 132f
Maskelyne, N, 197
Maxwellian relaxation time, 203
Mayer, C, 182
Mercury, 8, 11ff, 31, 76, 99ff, 112, 131f, 139f, 144, 154ff, 160, 164, 176, 178, 187, 191, 202, 210
  atmosphere of, 108
  Caloris Basin, 105f
  climate, 104
  craters on, 104ff
  dimensions of, 101f
  internal structure, 107
  magnetic field, 107f
  mass of, 102
  orbit of, 99f
  rotation of, 103f
  sidereal year of, 100
  synodic year of, 100
Mesopotamia, 8
Meteor showers, 166ff
  radiant of, 166
  Andromedids, 166
  Draconids, 166
  Leonids, 166
  Perseids, 166
Meteorites, 162ff
  carbonaceous chondrites, 87, 157, 165

chondritic, 87, 165
Sikhote Alin meteorite, 164
Meteors, sporadic, 166
   orbits of, 167
Mohorovičić, A. 203
Month, 41, 97
   anomalistic, 42
   draconic, 42f
   sidereal, 41f
   synodic, 42f
Moon, 2, 4f, 7f, 10, 13f, 19, 21ff, 25ff, 29ff, 39ff, 43f, 47, 51ff, 76ff, 111ff, 125, 129, 132, 140, 144, 155ff, 160f, 164f, 167, 173f, 180, 187, 202, 210f
   age of, 87
   chronology of, 80ff
   conductivity of, 51
   distance to, 40f
   eclipses of, 42f, 72f
   environment of, 66ff
   exosphere of, 76f
   gravitational field of, 62, 77
   internal structure of, 46ff, 107
   librations of, 44f, 49f
   magnetic field of, 52
   'mascons', 20
   mass of, 45, 50, 87
   motion of, 41ff
   orbit of, 41f, 96
   origin of, 91ff
   radius of, 13, 46
   rotation of, 44
   shape of, 29f
   surface structure of, 74ff
   temperature of, internal, 51; surface, 68, 70ff
   Tranquillity Base, 69
   velocity of, 41
Moonquakes, 47ff
   'signatures' of, 48ff
Moroz, V, 182
Muller, P, 20

Neptune, 9ff, 99, 102
Newton, Isaac, 9, 40, 196f
Nicholson, S B, 182

Olbers, J G, 152f, 158
Öpik, E J, 131, 159

Paulus, Aemilius, 43
Perseus, King of Macedonia, 43
Peterson, J H, 170
Pettit, E, 117, 182
Phosphorus, 173
Photosynthesis, 212
Piazzi, G, 152
Pillars of Hercules, 1
Piotrowski, S L, 159
Planck's law, 71, 170, 182
Planetesimals, 161
Planets,
   major, 10
   minor, 10
   orbits of, 10
   outer, 111
   rings of, 10
   satellites of, 10
   terrestrial, 10f, 99
Pluto, 9, 99ff, 109f, 112
   climate, 110
   dimensions of, 101f
   mass of, 102
   orbit of, 100
   rotation of, 108
Poincaré, H, 40
Poynting–Robertson effect, 171
Prutkov, Kuzma, 7

Rayleigh scattering, 124
Reflectors, cube-corner, 41
Regiomontanus (J Müller), 43
Roche limit, 159
Rockets,
   Saturn C-5, 22f
   Titan–Centaur, 34

Saari, J, 73
Saros, 43
Saturn, 8f, 11f, 37, 155, 159f
   rings of, 159
Selene, 39
Shorn, R A, 182
Shorthill, R W, 73
Sinton, W M, 117, 182
Sjogren, W L, 20
Solar corona, 170
Solar system,
   age of, 87, 99, 171
   Copernican (heliocentric), 8f, 173
   Ptolemaic, 9, 173
   Tychonic, 174

Index

Solar wind, 51, 169, 193
Spacecraft,
    ALSEP, 27f, 77
    Apollo, 3ff, 16, 22, 24ff, 41, 46, 50ff, 58, 63f, 75ff, 83ff
    Eagle (Apollo 11 excursion module), 3, 69
    Explorer, 15f, 20, 31, 51f, 208
    Luna, 3, 13ff, 24, 30f, 41, 45, 52
    Lunokhod, 18f, 27
    Mariner, 31ff, 35, 101f, 104, 116f, 120, 124, 133ff, 145ff, 183, 185
    Mars, 32, 118, 148
    Orbiter, 15, 17, 19f, 45, 56, 74
    Pioneer, 13, 32, 36f, 168, 171
    Surveyor, 14ff, 18, 21, 34, 71
    Venera, 31ff, 183f, 189ff, 193
    Viking, 32ff, 111, 115, 117ff, 123f, 125ff, 140, 142, 145ff, 182
    Zond, 15f, 30, 32, 45
Störmer, C, 208
Strong, J, 117, 182
Sun, 7ff, 14, 44, 47, 68, 70f, 92, 100, 103, 110, 122, 151, 154, 156f, 159f, 166ff, 173f, 176f, 179f, 183f, 193, 195, 208
    eclipse of, 14, 43, 60, 74, 78, 111, 170

Tanit of Salammbo, 39
Thucydides, 43
Tidal friction, 95, 107, 147, 180, 216

Tolstoy, Alexei, 7
Tombaugh, C, 99
Turkevich, A, 14

Ultima Thule, 1
Uranus, 9, 11f, 102, 152, 155, 159f
    mass of, 11
Urey, H C, 93

Venus, 8, 12, 31, 33, 40, 102, 107, 112, 173ff, 202, 207
    atmosphere of, 31, 180ff, 192ff
    composition of, 186
    axial rotation, 177ff
    clouds of, 173, 178, 184ff
    internal structure of, 189, 191
    magnetic field of, 191
    mass of, 176
    orbit of, 174
    orbital period, 179f
    phases of, 173ff
    surface of, 31, 33, 189f
    temperature of, 182f
    tidal coupling, 180

Widmannstätten figures, 164
Wildt, R, 158

Zodiacal cloud, 10, 169ff
Zodiacal light, 167